Sustainable Development Goals Series

The **Sustainable Development Goals Series** is Springer Nature's inaugural cross-imprint book series that addresses and supports the United Nations' seventeen Sustainable Development Goals. The series fosters comprehensive research focused on these global targets and endeavours to address some of society's greatest grand challenges. The SDGs are inherently multidisciplinary, and they bring people working across different fields together and working towards a common goal. In this spirit, the Sustainable Development Goals series is the first at Springer Nature to publish books under both the Springer and Palgrave Macmillan imprints, bringing the strengths of our imprints together.

The Sustainable Development Goals Series is organized into eighteen subseries: one subseries based around each of the seventeen respective Sustainable Development Goals, and an eighteenth subseries, "Connecting the Goals," which serves as a home for volumes addressing multiple goals or studying the SDGs as a whole. Each subseries is guided by an expert Subseries Advisor with years or decades of experience studying and addressing core components of their respective Goal.

The SDG Series has a remit as broad as the SDGs themselves, and contributions are welcome from scientists, academics, policymakers, and researchers working in fields related to any of the seventeen goals. If you are interested in contributing a monograph or curated volume to the series, please contact the Publishers: Zachary Romano [Springer; zachary.romano@springer.com] and Rachael Ballard [Palgrave Macmillan; rachael.ballard@palgrave.com].

More information about this series at https://link.springer.com/bookseries/15486

Paul Vare • Nadia Lausselet
Marco Rieckmann
Editors

Competences in Education for Sustainable Development

Critical Perspectives

Springer

Editors
Paul Vare
School of Education and Humanities
University of Gloucestershire
Cheltenham, UK

Nadia Lausselet
University of Teacher Education Vaud
Lausanne, Switzerland

Marco Rieckmann
Faculty of Education and Social Sciences
University of Vechta
Vechta, Germany

The content of this publication has not been approved by the United Nations and does not reflect the views of the United Nations or its officials or Member States.

Color wheel and icons: From https://www.un.org/sustainabledevelopment/, Copyright © 2020 United Nations. Used with the permission of the United Nations.

ISSN 2523-3084 ISSN 2523-3092 (electronic)
Sustainable Development Goals Series
ISBN 978-3-030-91054-9 ISBN 978-3-030-91055-6 (eBook)
https://doi.org/10.1007/978-3-030-91055-6

This Springer imprint is published by the registered company Springer Nature Switzerland AG
The registered company address is: Gewerbestrasse 11, 6330 Cham, Switzerland

To Andre; your dedication inspires us.

Acknowledgements

Firstly, we gratefully acknowledge the support of the European Union and the Erasmus+ programme that supported two phases of A Rounder Sense of Purpose; specifically, project number 2018-1-UK01-KA203-048121 made this book possible.

As ever, many people have contributed to the production of this volume. As editors we would like to say a special thank you to all our contributing authors who gave freely of their time to share their work and their insights with us.

We are indebted to the series editors and production team at Springer for bearing with us while we all waited for 'just one last chapter' and for all the work that followed after the writing was done.

Finally, and by no means least, we would like to thank our loved ones who have tolerated, largely with good grace, the havoc on family life that projects like these inevitably wreak.

Contents

Part I Conceptualising Competences

1 **Introduction** 3
Paul Vare, Marco Rieckmann, and Nadia Lausselet

2 **The Competence Turn** 11
Paul Vare

3 **Educators' Competence Frameworks in Education for Sustainable Development** 19
Marco Rieckmann and Matthias Barth

4 **What Do Key Competencies in Sustainability Offer and How to Use Them** 27
Arnim Wiek and Aaron Redman

5 **A Rounder Sense of Purpose: Competences for Educators in Search of Transformation** 35
Rick Millican

6 **On the Educational Difference Between Being Able and Being Willing** 45
Kerry Shephard

7 **Complexity and Criticality in Relation to ESD Competences** 53
Alain Pache and Sybille Rouiller

8 **Post-Normal Science, Transdisciplinarity, and Uncertainty in Relation to Educators' Competences: A Conversation with Silvio Funtowicz** 61
Francesca Farioli, Silvio Funtowicz, and Michela Mayer

9 **ESD Competences for Deep Quality in Education** 69
Chrysanthi Kadji-Beltran and Aravella Zachariou

Part II Integrating Competences

10 **The Highs and Lows of Working ESD Competences into HEI Frameworks** 79
Isabel Ruiz-Mallén, María Heras, Ramon Ribera-Fumaz, Hug March, and Andrea Corres

11 **Breaking the Mold: Educators as Agents of Change** 85
Francesca Farioli and Michela Mayer

12 **Framing the Frames: Integrating an ESD Approach Into an Existing National Framework** 93
Mónika Réti, Edit Lippai, and Márk Nemes

13 **The Rise of Competence-Based Learning in the Netherlands and Its Impact on the Implementation of ESD at Marnix Academie** 103
Gerben de Vries, Stella van der Wal-Maris, André de Hamer, and Carlien Nijdam

14 **Working on ESD Teacher Competences in Switzerland: A Case Study at the University of Teacher Education Vaud** . . . 113
Alain Pache and Nadia Lausselet

15 **Development and Assessment of ESD Competences: Staff Training at the University of Vechta** 121
Lukas Scherak and Marco Rieckmann

16 **ESD Competences and Teacher Educators' Professional Development: Findings and Implications of the Cyprus Example** 129
Aravella Zachariou and Chrysanthi Kadji-Beltran

Part III Pedagogy and Competences

17 **Connections Between Sustainable Development Competences and Pedagogical Approaches** 139
Rodrigo Lozano and Maria Barreiro-Gen

18 **Art and Connectedness within Sustainability: Educating Through Aesthetic Pedagogies** 145
María Heras

19 **Gamification for Sustainability: A Possible Renaissance of Role-Playing Game Mechanics in Pedagogy** 159
Mónika Réti, Edit Lippai, and Márk Nemes

20 **Outdoor and Sustainability Education: How to Link and Implement Them in Teacher Education? An Empirical Perspective** 167
Nadia Lausselet and Ismaël Zosso

21 Assessing Sustainability Competences: A Discussion on *What* and *How* . . . 175
Francesca Farioli, Michela Mayer, Rick Millican, Aaron Redman, and Paul Vare

Part IV Conclusions

22 Lessons Learned and Future Research Directions in Educating for Sustainability Competencies . . . 185
Ingrid Mulà, Gisela Cebrián, and Mercè Junyent

23 Closing Thoughts: The Role of Educator Competences for ESD in Turbulent Times . . . 195
Paul Vare, Nadia Lausselet, and Marco Rieckmann

Index . . . 201

Contributors

Maria Barreiro-Gen Faculty of Engineering and Sustainable Development, University of Gävle, Gävle, Sweden

Matthias Barth Eberswalde University for Sustainable Development, Eberswalde, Germany

Gisela Cebrián Department of Pedagogy, Universitat Rovira i Virgili, Tarragona, Spain

Andrea Corres Internet Interdisciplinary Institute (IN3), Universitat Oberta de Catalunya, Barcelona, Spain

André de Hamer DuurzamePABO, Sint Kruis, The Netherlands

Gerben de Vries DuurzamePABO, Sint Kruis, The Netherlands

Francesca Farioli Italian Association for Sustainability Science (IASS), Rome, Italy

Silvio Funtowicz Centre for the Study of the Sciences and the Humanities (SVT), University of Bergen (UiB), Bergen, Norway

María Heras Internet Interdisciplinary Institute (IN3), Universitat Oberta de Catalunya, Barcelona, Spain

Mercè Junyent Research Group COMPLEX, Universitat Autònoma de Barcelona, Barcelona, Spain

Chrysanthi Kadji-Beltran School of Education, Department of Education, Frederick University, Nicosia, Cyprus

Nadia Lausselet International Research Laboratory for Education for Sustainable development (LirEDD), University of Teacher Education State of Vaud, Lausanne, Switzerland

Edit Lippai Hungarian Research Teachers Association, Budapest, Hungary

Rodrigo Lozano Faculty of Engineering and Sustainable Development, University of Gävle, Gävle, Sweden

Organisational Sustainability, Ltd., Cardiff, UK

Hug March Internet Interdisciplinary Institute (IN3), Universitat Oberta de Catalunya, Barcelona, Spain

Estudis d'Economia i Empresa, Universitat Oberta de Catalunya, Barcelona, Spain

Michela Mayer Italian Association for Sustainability Science (IASS), Rome, Italy

Rick Millican School of Education and Humanities, University of Gloucestershire, Cheltenham, UK

Ingrid Mulà Research Group on Science and Environmental Education, Universitat de Girona, Girona, Spain

Márk Nemes Hungarian Research Teachers Association, Budapest, Hungary

Carlien Nijdam Hogeschool Zeeland, Vlissingen, The Netherlands

Alain Pache International Research Laboratory for Education for Sustainable development (LirEDD), University of Teacher Education State of Vaud, Lausanne, Switzerland

Aaron Redman School of Sustainability, College of Global Futures, Arizona State University, Tempe, AZ, USA

Mónika Réti Hungarian Research Teachers Association, Budapest, Hungary

Ramon Ribera-Fumaz Internet Interdisciplinary Institute (IN3), Universitat Oberta de Catalunya, Barcelona, Spain

Marco Rieckmann Department of Education, Faculty of Education and Social Sciences, University of Vechta, Vechta, Germany

Sybille Rouiller Didactics of Human and Social Sciences Unit, University of Teacher Education State of Vaud, Lausanne, Switzerland

Isabel Ruiz-Mallén Internet Interdisciplinary Institute (IN3), Universitat Oberta de Catalunya, Barcelona, Spain

Lukas Scherak Department of Education, Faculty of Education and Social Sciences, University of Vechta, Vechta, Germany

Kerry Shephard Higher Education Development Centre, University of Otago, Dunedin, New Zealand

Stella Van der Wal-Maris Marnix Academie, Utrecht, The Netherlands

Paul Vare School of Education and Humanities, University of Gloucestershire, Cheltenham, UK

Arnim Wiek School of Sustainability, College of Global Futures, Arizona State University, Tempe, AZ, USA

Center for Global Sustainability and Cultural Transformation, Arizona State University, AZ, USA

Aravella Zachariou Unit of EESD/CPI, Government of Cyprus and Frederick University, Nicosia, Cyprus

Ismaël Zosso Outdoor Competence Centre, University of Teacher Education State of Vaud, Lausanne, Switzerland

Abbreviations

CBET	Competence-Based Education and Training
CBL	Competence-Based Learning
CI	Controversial Issues
CK	Content Knowledge
DeSeCo	Definition and Selection of Competencies
EFCA	Educating Future Change Agents
EfS	Education for Sustainability
EG	Educational Games
ESD	Education for Sustainable Development
ESE	Environmental and Sustainability Education
G4L	Games for Learning
HESD	Higher Education for Sustainable Development
IASS	Italian Association for Sustainability Science
IEEP	International Environmental Education Programme
IQ	Intelligence Quotient
ISCED	International Standard Classification of Education
LfS	Learning for Sustainability
MA	Marnix Academie
OE	Outdoor Education
OED	Oxford English Dictionary
PCK	Pedagogical Content Knowledge
PD	Professional Development
PISA	Programme for International Student Assessment
PLLOs	Program-Level Learning Objectives
PNS	Post-Normal Science
QE	Quality Education
RPGs	Roleplaying Games
RSP	A Rounder Sense of Purpose
SD	Sustainable Development
SDGs	Sustainable Development Goals
SE	Sustainability Education
SNPA	National System for Environmental Protection
TOE	Transformative Outdoor Education
TTIs	Teacher Training Institutes
UN	United Nations
UNECE	United Nations Economic Commission for Europe
UNESCO	United Nations Educational, Scientific and Cultural Organisation
UOC	Universitat Oberta de Catalunya
WSA	Whole School Approach

Part I

Conceptualising Competences

1 Introduction

Paul Vare, Marco Rieckmann, and Nadia Lausselet

Abstract

This introductory chapter introduces the case for competences in education for sustainable development before going on to explore the concept of competence itself. Three types of competence are identified: a pragmatic understanding comprising a focused set of skills, knowledge and attitudes; a broader concept that aligns with the German *Bildung* tradition; an innate quality or potential that lies within a given organism, institution or system. These and related concepts are illustrated in a novel concept map that shows how each relates to the other. The chapter goes on to outline the structure of the book, providing a brief overview of each chapter in turn.

Keywords

Competences · Education for sustainable development · Concept mapping · Educator competences

P. Vare (✉)
University of Gloucestershire, Cheltenham, UK
e-mail: pvare@glos.ac.uk

M. Rieckmann
University of Vechta, Vechta, Germany

N. Lausselet
University of Teacher Education, Vaud, Lausanne, Switzerland

Sustainable Development and the Role of Education

There is no shortage of evidence to show that environmental and social challenges threaten our well-being if not the habitability of the Earth. If, as Nobel laureate Paul Crutzen proposed, we are entering the Anthropocene—the age of human beings—this would be the first time in Earth's history that a shift in geological epoch has been driven by a single species. Perhaps most strikingly of all, we *know* that we are doing this.

To address this situation, the concept of sustainable development has been operationalised in the form of internationally accepted Sustainable Development Goals (SDGs) adopted by the 193 countries of the United Nations General Assembly (UN 2015). The 17 SDGs, together with their associated 169 targets, present a formidable 'to do' list, all of which requires a good deal of learning. With this in mind, UNESCO (2017) has drafted a set of learning objectives linked to each of the SDGs. Leaving aside the larger issue of whether this is truly the 'future we want' (Kopnina 2017), this still leaves the question: what kind of qualities does an educator need to have in order to provide effective learning for sustainability? And, having established these qualities, how might these be acquired?

It is the aim of this book to look critically at these questions in light of efforts that have been made over recent decades to answer them. These

P. Vare et al. (eds.), *Competences in Education for Sustainable Development*, Sustainable Development Goals Series, https://doi.org/10.1007/978-3-030-91055-6_1

efforts have tended to focus on defining competences, whether for learners (e.g., Wiek et al. 2011; Rieckmann 2018; Brundiers et al. 2021) or for educators (e.g., Sleurs 2008; UNECE 2012; Vare et al. 2019). While this book focuses on the competences of educators, we acknowledge that all this work reflects a wider trend in education that we call the Competence Turn (see Chap. 2); that is, a trend to focus on the *outcomes* of education rather than inputs such as curriculum content.

What's in a Name?

Before we investigate the concept of competence, we should say a word about the field of education in which our work is located. We have already used the term education for sustainable development (ESD) in our title and learning for sustainability (LfS) above. Read on and you will soon encounter environmental and sustainability education (ESE) and even sustainability education (SE). Each of these terms has its supporters and detractors but this is not the place to rehearse those arguments.[1] Most authors in this book use ESD as the long-standing, internationally accepted term, others prefer ESE which foregrounds the environment, which after all is the ultimate 'bottom line'. We ask that you do not read too much into these differences across the chapters although we accept that words *do* make a difference.

As for the term competence, there appears to be no single, definitive understanding of this. Firstly, there is the distinction between *competence* and *competency*. For Hyland (1994) competence is the broader term, he sees it as a capacity to fulfil a certain role or procedure, whereas competency is narrower and refers to a specific skill or ability. In the USA, competency has been seen as the broader term compared to the more specific 'learning outcome' (Wilke et al., 1987). Another quirk of language is that the plural form of competency (competencies) is often used, even when discussing the singular form 'competence', for example, by Klieme et al. (2008) and Wiek et al. (2011). All of which suggests that we should not attach too much importance to the decision to use competence or competency. After all, Mula et al. (Chap. 22) reassure us that both terms translate as *competencia* in Spanish, while *Kompetenz* does the job in German.

One reasonable justification for any apparent grammatical oversight is that competence can be understood as a quality, an uncountable noun, as well as a specific attribute that might be denoted by having *a* competence. This highlights the need to clarify which concept of competence one is using.

Klieme et al. (2008) identify three distinct theoretical concepts of competence: Firstly, drawing on the pragmatic psychological tradition is a construct promoted by David McClelland who, in 1973, proposed competence diagnostics as an alternative to traditional intelligence tests. This is categorically not a trait, rather it is something to be acquired with practice. This concept of competence can be summarised as 'context-specific dispositions for achievement that can be acquired through learning' (Klieme et al. 2008 p. 7). Accepting that the use of *dispositions* here suggests 'readiness' rather than personality traits, which is how the term is commonly used in English-speaking countries, this is perhaps closest to the popular understanding of competence as it is used in the UK, where the term is often conflated with skills or a 'skill set', a view popularised with the growth of National Vocational Qualifications in the 1980s. We call this Type 1 in our concept map (Fig. 1.1).

A second use of the term, proposed by McClelland's contemporary, the German Heinrich Roth, aims to bridge the divide between the tradition of *Bildung*, an open term encompassing the formation of one's personality, and the notion of qualification as applied to vocational learning. For Roth, competence is aligned with an individual's increasing maturity and autonomy. This may help to explain the broad definition of competence in German-speaking countries, which embraces motives as well as

[1] For a useful discussion on terminology in this field, see Sterling (2010).

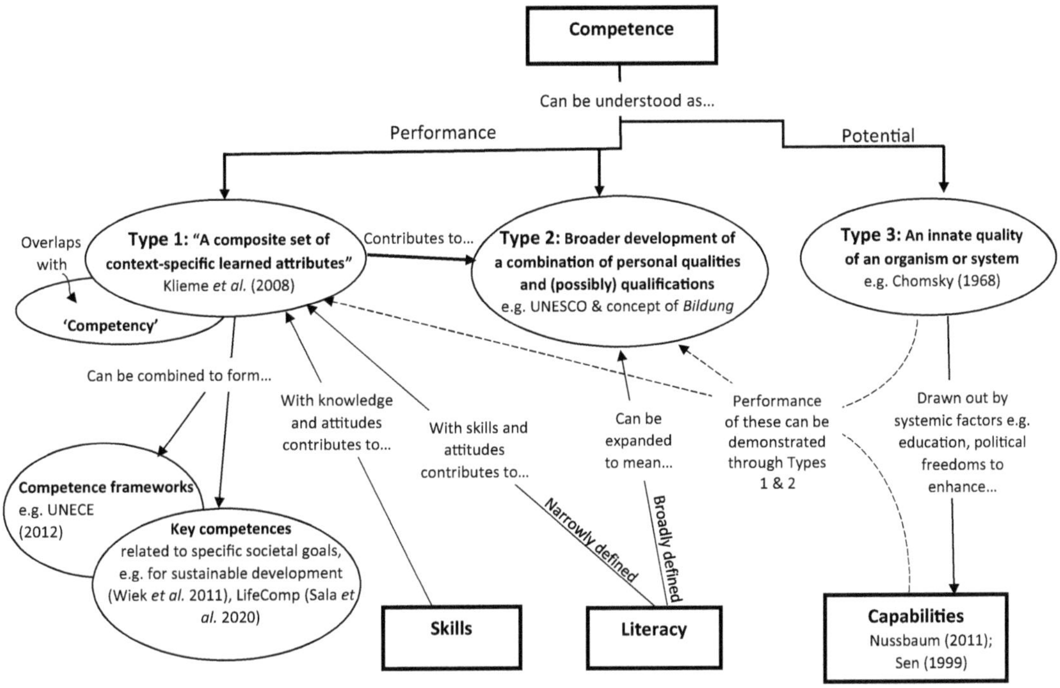

Fig. 1.1 A competence concept map

attitudes (Rieckmann 2012). With this broader understanding of the concept, it is unsurprising that it was in Germany that the ESD competence debate developed the term 'Gestaltungskompetenz' or transformation competence, described earlier by de Haan (2006) as 'shaping competence':

> Those who possess this competence can help, through their active participation in society, to modify and shape the future *of* society, and to guide its social, economic, technological and ecological changes along the lines of sustainable development. (Ibid, p. 22 – italics as in the original)

While this is specifically concerned with the capacity for action and problem solving, it comprises a range of different sub-competences such as foresighted thinking, interdisciplinary work, participatory skills, planning, implementation skills and the capacity for empathy. Crucially, the list includes self-motivation, a capacity that speaks to who we are as much as what we might know or do.

Several authors in this book adopt a broad-based definition of competence which would fit the pragmatic Type 1 while reflecting elements of this broader Type 2; this is shown on the concept map by an arrow indicating that Type 1 can be seen as a component of Type 2. UNESCO's definition that includes cognitive, affective, volitional and motivational elements (quoted more fully in Chap. 3) is a good example of this inclusive definition. While such an approach embraces an impressive range of attributes, it does create a wide umbrella term that can pose problems at the implementation stage when defining pedagogical approaches or assessment criteria.

A third concept (Type 3) of competence is where it refers to a general ability or trait, a definition favoured by Noam Chomsky (1968) to describe a communicative competence or facility for language. This is a communal property not something that has much meaning at the individual level and is therefore quite distinct from performance. *Homo sapiens*, for example, have a language competence that *Nematoda* do not. This idea of potential that may be realised with a favourable set of conditions, be that education, political freedoms or other resources, is similar to the notion of capabilities as outlined by Sen (1999) and Nussbaum (2011). This view avoids the deficit model that an effort to teach competences can imply.

The specific concept of *key competences* (Weinert 2001; Rychen 2004) refers to constellations of these competences that facilitate an individual's performance in a particular domain, for example, in digital literacy. The competence projects of the European Union (see Chap. 2) fall into this category as do the sustainability competences developed by Wiek et al. (2011). A Type 1 competence lends itself well to this arrangement; a number of skills can be applied in combination to give what might be thought of as a broader competence in the sense that Roth proposed (Type 2). We cannot assume that an elision of Type 1 and Type 2 competences is always possible given the different traditions that underpin them. Indeed, the lack of conceptual clarity around competence leads Willbergh (2015) to call for the term to be abandoned altogether in favour of *Bildung* but this term is not without its own theoretical fault lines as explored in detail by Herdt (2019).

In some parts of the world, the umbrella nature of the term competence is avoided altogether; a literature review by Sterling et al. (2017) noted how:

> ...authors from Australia prefer the terms 'skills' and 'capabilities', and authors from the UK 'skills' and 'literacy'. Surprisingly, more than half of the articles (57%) give no definition of the terms used. (Sterling et al. 2017, p. 162)

Given that these alternative terms are applied where others might use competences, we have included them on the concept map (Fig. 1.1) to show how they could relate to the three competence types. Literacy is linked to both Type 1 and Type 2 depending on the definition used. A commonplace or narrow understanding of literacy refers to an ability to read, write and apply these skills freely for one's own purpose. Literacy has also been used in relation to ecology (Orr 1992) and sustainable development (Stibbe 2009) in ways that embrace a far wider set of attributes that extend beyond Type 1 competence and contribute more readily to Type 2.

Within this volume, as with the various permutation of ESD, ESE and so forth, we have not demanded that our various contributors adhere to one type or even one spelling of competence, rather we wish to demonstrate how the term is used differently in a variety of contexts. It is for you, our reader, to decide on the extent to which you feel that they are all talking about the same thing.

The Structure and Contents of This Book

This brings us to the contributions themselves. Our volume is divided into four parts:

Part I introduces a number of conceptual debates around competences beginning with a brief historical account of 'The Competence Turn' in Chap. 2 by Paul Vare. This includes warnings about some of the hazards inherent in the enthusiastic adoption of a competence-based approach.

In Chap. 3, Marco Rieckmann and Matthias Barth remind us that the different ESD competence frameworks currently available are focused on specific target groups, something that is often overlooked in the literature. They call for the training of educators to be more structured around ESD competence frameworks although programme managers need to decide how far they should focus on the development of professional competences for ESD and general sustainability competences.

Arnim Wiek and Aaron Redman provide guidance in Chap. 4 to scholars, educators and administrators on how to improve competence-based educational practice and better contribute to advancing sustainability, by developing sustainability competences. They highlight flaws in current practices such as insufficient coverage and integration of competence content and the need for closer alignment between competences, pedagogies and assessment—both of competence development and professional success.

Many contributors to this book have been partners in *A Rounder Sense of Purpose* (RSP), a European Union-funded project that has developed a broad ESD competence framework for use by educators in any context. In Chap. 5, Rick Millican introduces the RSP framework, explains its structure and outlines some of the challenges faced before discussing some possible next steps.

As discussed above, competence is something of an umbrella term and in Chap. 6, Kerry Shephard delves deeply into the question of whether a single concept can embrace both the ability to do something and the willingness to put that ability into practice. Along the way he expands on the historical context of what we have called the Competence Turn.

In Chap. 7, Alain Pache and Sybille Rouiller focus on two core competences for ESD: critical and complex thinking. Their work highlights why it is important to address these concepts in a sophisticated manner as certain types of critical thinking can become counterproductive.

Chap. 8 takes the form of an interview or conversation between Silvio Funtowicz, who proposed the concept of Post-normal Science in 1993, and colleagues from the Italian Association of Sustainability Science, Francesca Farioli and Michela Mayer. Both ESD and Post-Normal Science recognise the need to dwell increasingly on the notion of uncertainty. This leads to their conversation underlining the importance of the co-creation of knowledge where errors or ignorance should no longer be seen as problems to be avoided but lessons to be learned and on which, even partially, we base our decisions.

The final contribution to Part I by Chrysanthi Kadji-Beltran and Aravella Zachariou represents one of the first attempts to link quality education to ESD competences. Their discussion in Chap. 9 is timely in the way that it presents the importance of quality education in opposition to instrumentalism and mere quantification. While stressing the importance of reforming educational systems they acknowledge the complexity of the task in terms of context and interconnectedness.

Part II focuses on the integration of ESD competence frameworks into various settings in different countries and at different scales. Each chapter comprises a case study, the first of which, presented by Isabel Ruiz-Mallén, María Heras, Ramon Ribera-Fumaz, Hug March and Andrea Corres, discusses the challenge of integrating ESD competences into a Spanish higher education institution's pre-existing competence framework. As they explain in Chap. 10, much rests on staff training and aligning the guiding principles of the institution with sustainable development.

While ESD emphasises the role of learners as change agents, Chap. 11 discusses the role of educators as change agents themselves. In this case, Francesca Farioli and Michela Mayer chart the progress of educators in developing this sensibility in three different programmes for non-formal and formal settings in Italy.

We zoom out to cover the national scale in Chap. 12. Mónika Réti, Edit Lippai and Márk Nemes review the multi-stakeholder process involved in integrating an ESD competence framework within an existing general framework used for assessing in-service teachers' competences in Hungary. They explore the structural challenges and pose the question of whether it is possible to deconstruct and rebuild a framework without losing its key values and ideals.

Institutions do not always have to go it alone as Gerben de Vries, Stella van der Wal-Maris, André de Hamer and Carlien Nijdam explain in Chap. 13. Marnix Academie, a teacher training institute in Utrecht, The Netherlands, had the support of Dutch NGO, DuurzamePABO as it set out on a long journey of cultural change. They highlight other external influences such as the UN Sustainable Development Goals and membership of UNESCO but also emphasise the painstaking nature of a slow, people-orientated, culture-focussed process that demands a multi-level, whole institution approach.

The need for making a whole institution approach more explicit is also raised in Chap. 14 where Alain Pache and Nadia Lausselet present the attempt made by the largest teacher education institution in French-speaking Switzerland (HEP Vaud) to align more sustainable learning environments with new ESD courses that have been launched working on ESD competences. They show the critical importance of external support with no time being lost in building on a favourable political environment. Even within this supportive context, examination of one course has shown the difficulty that student teachers face in imagining some of the broader concepts presented by the competence framework.

Chap. 15 presents the implementation of an ESD academic staff training programme at the University of Vechta (Germany) with a particular focus on the development and assessment of ESD competences. Lukas Scherak and Marco Rieckmann demonstrate the importance of an integrative approach when considering the ESD competences and highlight the challenge of raising enthusiasm for ESD among teachers who do not yet have an affinity for it.

Aravella Zachariou and Chrysanthi Kadji-Beltran provide the last case study in Part II. In Chap. 16 they outline research conducted among ESD educators in Cyprus who have been trained using the Rounder Sense of Purpose (RSP) competence framework. Despite a positive response to the RSP model, it needs time to be 'intellectually digested' and integrated critically. The authors make the important point that the most effective professional development comes not from the implementation of a particular set of 'best practices' but from their careful adaptation to the specific context in which they are to be applied.

Part III is concerned with a range of pedagogical approaches for developing competences in ESD and sustainable development itself. Chap. 17 opens a review by Rodrigo Lozano and Maria Barreiro-Gen of the connections between sustainability competences and the pedagogical approaches used to develop them. The authors suggest that a combination of pedagogical approaches is needed to better develop sustainability competences. Such a combination should encourage teachers and their students to challenge traditional mono-disciplinary and siloed approaches in order to provide solutions to complex, multi-level and wicked problems.

In Chap. 18 María Heras introduces us to 'aesthetic learning' as a possible way to develop a range of sustainability competences. Her review of art-based approaches in the literature reveals methods by which educators and learners can sense their ways of being in the world and promote an awareness of interconnectedness while exploring their capacity 'to be, to change, to care'. Although not prescriptive, this approach suggests a transformative learning potential that can be tapped by sustainability educators willing to engage in the mystery and open-ended nature of aesthetic experience.

Gamification is the focus of Chap. 19 by Mónika Réti, Edit Lippai and Márk Nemes. Their claim is that by adapting gamified methods that speak to upcoming generations that have been socialised in the media-landscape of the Twenty-first Century, educators may better develop sustainability-related competences in their learners. While taking care to differentiate between educational gaming and games-for-learning, the authors show how gamified lessons key in to our neurobiology so effectively that they can raise serious ethical concerns.

Outdoor education (OE) may be a more familiar context for ESD but in Chap. 20, Nadia Lausselet and Ismaël Zosso show that OE need not necessarily contribute to sustainability education. Their model of transformative OE combined with action research conducted alongside two teacher education programmes with an explicit focus on ESD competences provides insights into promising avenues and some possible limitations alongside more general support for transformative ESD.

Chap. 21 takes the form of a three-way conversation about the assessment of competences. With contributions from Aaron Redman in the USA, Francesca Farioli and Michela Mayer in Italy and Rick Millican and Paul Vare in the UK, these groups of authors share insights from their respective research endeavours before acknowledging that there is no 'perfect' method and that ultimately, assessment of something as complex as ESD competences must involve a range of different methods and approaches to assessment.

Part IV is our shortest section and provides a reflective overview of the field. In Chap. 22 Ingrid Mulà, Gisela Cebrián and Mercè Junyent survey the work done on ESD competences and echo some of the critiques made earlier in the book including a lack of conceptual clarity and the need to contextualise competences. Looking ahead, they note how the COVID-19 pandemic has accelerated the use of distance learning via digital platforms and they point to important work still to be done on how best to support learners in the development of their sustainability competences through online teaching and learning.

Our concluding Chap. 23 returns to the theme of the pandemic. As editors we look at how some of the themes that recur in this book might evolve in future and ultimately, we hope, contribute to our collective efforts to learn our way forward into a more sustainable world.

References

Brundiers, K., Barth, M., Cebrián, G., Cohen, M., Diaz, L., Doucette-Remington, S., Dripps, W., Habron, G., Harré, N., Jarchow, M., Losch, K., Michel, J., Mochizuki, Y., Rieckmann, M., Parnell, R., Walker, P. & Zint, M. (2021). Key competencies in sustainability in higher education—toward an agreed-upon reference framework. *Sustainability Science*, 16(1), 13–29. DOI: https://doi.org/10.1007/s11625-020-00838-2

Chomsky, N. (1968). *Language and Mind*. New York: Harcourt, Brace & World, Inc.

Haan G. de (2006). The BLK '21' programme in Germany: a 'Gestaltungskompetenz'-based model for Education for Sustainable Development. *Environmental Education Research, 12:1, 19–32.*

Herdt, J. A. (2019) *Forming Humanity: Redeeming the German Bildung Tradition*, Chicago: University of Chicago Press

Hyland, T. (1994) *Competence, Education and NVQs: Dissenting perspectives*. London: Cassell Education.

Klieme, E., Hartig, J., & Rauch, D. (2008). The concept of competence in educational contexts. In J. Hartig, E. Klieme, & D. Leutner (Eds.), *Assessment of competencies in educational contexts (pp. 3–22). Hogrefe & Huber Publishers.*

Kopnina, H. (2017) Future scenarios for sustainability education: the future we want? In Envisioning futures for environmental and sustainability education in Corcoran, P. B., Weakland, J. P. & Wals A. E. J. (Eds.) *Envisioning futures for environmental and sustainability education*. Wageningen, NL: Wageningen Academic Publishers.

Nussbaum, M. (2011). *Creating capabilities: The human development approach*. Harvard University Press.

Orr, D. (1992) *Ecological Literacy: Education and Transition to a Postmodern World*. Albany: SUNY

Rieckmann, M. (2012). Future-oriented higher education: Which key competencies should be fostered through university teaching and learning? *Futures, 44, 127–135*

Rieckmann, M. (2018). Chapter 2 - Learning to transform the world: key competencies in ESD. In A. Leicht, J. Heiss & W. J. Byun (Eds.), *Education on the move. Issues and trends in education for sustainable development (pp. 39–59). United Nations Educational, Scientific and Cultural Organization*

Rychen, D. S. (2004). Key competencies for all: an overarching conceptual frame of reference. In D. S. Rychen, D. S. & Tiana, A. (Eds.), *Developing Key Competencies in Education: Some Lessons from International and National Experience* (pp. 5–34). UNESCO, International Bureau of Education.

Sala, A., Punie, Y., Garkov, V. & Cabrera Giraldez, M. (2020) *LifeComp: The European Framework for Personal, Social and Learning to Learn Key Competence, EUR 30246 EN, Publications Office of the European Union, Luxembourg, ISBN 978-92-76-19417-0, doi:*https://doi.org/10.2760/922681, JRC120911

Sen, A. (1999) *Development as freedom*. Oxford: OUP

Sleurs, W. (2008) *Competencies for ESD (Education for Sustainable Development) Teachers: A Framework to Integrate ESD in the Curriculum of Teacher Training Institutes; Curriculum, Sustainable Development, Competences, Teacher Training (CSCT); Comenius 2.1 Project; UN: Brussels, Belgium, 2008; Available online:* https://www.unece.org/fileadmin/DAM/env/esd/inf.meeting.docs/EGonInd/8mtg/CSCT%20Handbook_Extract.pdf (accessed on 27 November 2020).

Sterling, S. (2010). Living in the Earth: Towards an Education for Our Time. *Journal of Education for Sustainable Development*, 4(2), 213–218. DOI: https://doi.org/10.1177/097340821000400208

Sterling, S., Glasser, H., Rieckmann, M. & Warwick P. (2017) "More than scaling up": a critical and practical inquiry into operationalizing sustainability competencies In Envisioning futures for environmental and sustainability education in Corcoran, P. B., Weakland, J. P. & Wals A. E. J. (Eds.) *Envisioning futures for environmental and sustainability education*. Wageningen, NL: Wageningen Academic Publishers.

Stibbe, A. (ed.) (2009) *The handbook of sustainability literacy*. Dartington: Green Books

UN – United Nations (2015) Resolution adopted by the General Assembly on 25 September 2015, *Transforming our world: the 2030 Agenda for Sustainable Development*. New York: UN https://en.wikipedia.org/wiki/File:N1529189.pdf

UNECE (2012) *Learning for the Future: Competences in Education for Sustainable Development*, Geneva: United Nations. https://unece.org/fileadmin/DAM/env/esd/ESD_Publications/Competences_Publication.pdf

UNESCO – United Nations Educational, Scientific and Cultural Organization (2017). *Education for Sustainable Development Goals. Learning Objectives*. Paris: UNESCO. http://unesdoc.unesco.org/images/0024/002474/247444e.pdf

Vare, P.; Arro, G.; de Hamer, A.; Del Gobbo, G.; de Vries, G.; Farioli, F.; Kadji-Beltran, C.; Kangur, M.; Mayer,

M.; Millican, R.; Nijdam, C.; Réti, M.; Zachariou, A. (2019) Devising a Competence-Based Training Program for Educators of Sustainable Development: Lessons Learned. *Sustainability, 11, 1890.* https://www.mdpi.com/2071-1050/11/7/1890

Weinert, F. E. (2001) Concept of competence: a conceptual clarification. In Rychen, D. S. & Salganik, L. H. (Eds.), *Defining and selecting key competencies* (pp. 45–65). Seattle: Hogrefe & Huber Publishers.

Wiek, A.; Withycombe, L.; Redman, C.L. (2011) Key competences in sustainability: A reference framework for academic program development. *Sustainability Science*, 6, 203–218. https://link.springer.com/article/10.1007/s11625-011-0132-6

Wilke, R.J., Peyton, R.B. & Hungerford, H.R. (1987). Strategies for the training of teachers in environmental education. *International Environmental Education Programme; environmental education series No. 25. Paris: UNESCO-UNEP.*

Willbergh, I. (2015) The problems of 'competence' and alternatives from the Scandinavian perspective of Bildung. *Journal of Curriculum Studies*, 47, 334–354. https://doi-org.glos.idm.oclc.org/10.1080/00220272.2014.1002112

2 The Competence Turn

Paul Vare

Abstract

The Competence Turn marks a shift of attention away from what is taught, with a focus on curriculum content, to what is learned, i.e. the outcomes of education. It is a shift with some considerable history. This chapter explores the approach to curriculum that underpins the Competence Turn and charts the growing use of the term 'competence' in education since the mid-Twentieth Century, looking specifically at how it has been pressed into the service of environmental and sustainability education. The final section discusses some hazards that accompany this focus on competence. These can be summarised as: conceptual confusion; the centrality of context; impacts on pedagogy; distractions from fundamental issues concerning the wider purpose of education.

Keywords

Critique of competences · Environmental and sustainability education · Curriculum approaches · Purpose of education

P. Vare (✉)
University of Gloucestershire, Cheltenham, UK
e-mail: pvare@glos.ac.uk

Introduction

Competences are everywhere. At the time of writing (early 2021), the European Union has embarked on a mission to identify the competences that its citizens need in order to become more sustainable. This will be the fourth in a suite of key competence frameworks. The first of these deals with digital literacy, *Digicomp* (Punie et al. 2013), followed by competences to encourage entrepreneurial mindsets, *Entrecomp* (Bacigalupo et al. 2016), while the third framework addresses personal, social and 'learning to learn' competences, *LifeComp* (Sala et al. 2020). The concept of competence appears to have us in such a hegemonic grip that it is difficult to imagine a time when we did not think about education and its outcomes in terms of competences—or competencies (see Chap. 1 for a discussion on terminology).

This concern with competence marks a shift away from what is taught, with its focus on curriculum content, to what is learned—the outcomes of education. It is a shift with some considerable history. This chapter explores the approach to curriculum that underpins the Competence Turn, charts the growing use of the term in education since the mid-Twentieth Century and looks specifically at how it has been pressed into the service of environmental and sustainability education. The final section dis-

P. Vare et al. (eds.), *Competences in Education for Sustainable Development*, Sustainable Development Goals Series, https://doi.org/10.1007/978-3-030-91055-6_2

cusses some of the hazards that accompany this focus on competence.

The Rise of Competence-Based Education

Before the Second World War, the dominant mode of curriculum development was what Kelly (2009) terms *curriculum as content*. An early proponent of this was Franklin Bobbit, an American superintendent of schools who sought to deliver content in the form of manageable component parts that could be 'transmitted' by teachers. By the mid-Twentieth Century education planners were less concerned with specifying curriculum content and more interested in the uses to which a learner might put their learning—or the uses to which the learner might be put. This led to the planning of education to serve a predetermined outcome, what Kelly (2009) terms *curriculum as product*. In 1949, the American educationalist, Ralph Tyler, sought to define the purpose of education in terms of clear, behavioural (and thus measurable) objectives, guided by four questions:

- What educational purposes should a school seek to attain?
- What educational experiences can be provided that are likely to attain those purposes?
- How can those educational experiences be effectively organised?
- How can we determine whether those purposes are being attained?

- (Tyler 1949 in Lawton 1996, p. 19)

This approach reflects a Newtonian logic that views processes as predictable and manageable. It underpins what came to be known as competence-based education and training (CBET) which, as Shephard explains in Chap. 6, became popular in the USA largely as a response to the 'space race' and that nation's perceived deficits in technical education relative to the USSR. Indeed, for technical and vocational education, CBET presents a practical way forward; however, it has not stopped there. Outcomes are now 'used as criteria for the productivity of entire educational systems' (Klieme et al. 2008, p. 3); an example from The Netherlands in relation to teacher education is discussed in Chap. 13. While the term competence may still lack conceptual clarity (Chap. 1), this objective-based approach is founded on an absolutist epistemology, a view advocated as far back as Plato. While content-focused learning does not necessarily dictate what should be done with that content, a competence-based approach certainly does. The way in which this approach lends itself to measurable verification with apparent ease resonates with the all-pervasive managerialism that has come to characterise national education systems in the wake of neoliberal policy environments that have become a global phenomenon since the 1980s (Harvey 2005). It may be that this close fit between ideology and managerial expediency explains the popularity of competences as much as any inherent educational benefits that they confer, particularly given the concerns that it raises; something I return to later.

Competences in Environmental and Sustainability Education (ESE)

Given that the clear, linear logic as expressed above has helped competence-based education to become a widespread and enduring approach, it is little wonder that this way of thinking has found its way into environmental and sustainability education (ESE). There is a distinction to be made between sustainability competence frameworks that describe what all of us should learn and education for sustainability competence frameworks that set out attributes that *educators* need to have to be able to support the development of learners' sustainability competences. It was actually the second of these types of frameworks that appeared first of all in the form of an international agreement. This was the set of learning outcomes for educators as defined by the *Inter-governmental Conference on Environmental Education* in Tbilisi, Georgia (then USSR) in 1977. There was no talk of competences back then, instead the outcomes distinguished between *awareness* and *knowledge* with *skills* and *participation* also listed separately. The other category, *attitudes*, spoke of 'a set of values' and 'the motivation for actively participating in environmental improvement and protection' (UNESCO-UNEP 1978, p. 27). This avoided the clustering of outcomes under broad competences that might otherwise be difficult to comprehend but it did lead to formidable lists of outcomes.

Some 10 years later, in an effort to make all this intelligible to educators, the International Environmental Education Programme (IEEP) published a set of environmental education 'competencies' for teachers as part of its series of 'green books' (Wilke et al. 1987). The term 'competencies' is understood broadly here, even so, the lists are daunting. Sub-divided into four levels, they cover: *(1) Ecological foundations; (2) Conceptual awareness; (3) Investigation and evaluation; (4) Environmental action skills.*

Almost two decades elapsed before, in 2005, the United Nations Economic Commission for Europe (UNECE) published its Strategy for Education for Sustainable Development. This includes a key action area to 'develop the competence within the education sector to engage in ESD' (UNECE 2009, p. 21). Interestingly, this could be read as a Type 3 competence as defined in the concept map in Chap. 1, that is a system wide competence or capability rather than the performance of the individuals within it. That is not the way things developed. An early attempt to address this key action was made by the international organisation ENSI (Environment and School Initiatives), which developed the CSCT model, i.e. *C*urriculum, *S*ustainable development, *C*ompetences, *T*eacher training (Sleurs 2008). As the ENSI project was nearing completion, UNECE itself convened an expert group to define ESD competences for educators, which led to the development of 39 competences gathered under three broad headings: (a) holistic approach; (b) envisioning change; (c) achieving transformation (UNECE 2012).

Both the ENSI and UNECE models provide a valuable insight into the competences required by educators to promote learning for sustainability but neither model has been adopted widely. This is possibly because their level of detail, while helpful, makes them so unwieldy that they cannot be adapted readily into today's crowded teacher education programmes. Efforts to define the role and competences of educators of sustainability have continued apace; as discussed in Chap. 3, this includes the KOM-BiNE model (Rauch and Steiner 2013), the work of Bertschy et al. (2013), A Rounder Sense of Purpose (Vare et al. 2019) and the work of Timm and Barth (2021).

Meanwhile, competence frameworks for sustainability per se have also evolved; perhaps the earliest being the *Definition and Selection of Competencies* (DeSeCo) project established by the Paris-based Organisation for Economic Cooperation and Development (OECD 2002). This aimed to identify the 'competencies' necessary for individuals to confront the challenges of balancing economic growth with environmental sustainability and social equity. It continues to provide the foundation of the *OECD Learning Compass 2030*, a project that aims to help young people succeed in, and shape, their future by identifying what are termed 'transformative competencies' (Rychen 2019). Interestingly, the notion of 'shaping the future' is captured by the German term *Gestaltungskomptenz* (de Haan 2006), the discourse on which informed an influential paper on sustainability key competencies in higher education (Barth et al. 2007). Another highly influential set of key competencies in sustainability, developed originally as learning outcomes of sustainability science students (Wiek et al. 2011; Chap. 4) went on to inform UNESCO's key competencies in education for sustainable development (Rieckmann 2018) as well as a recent international Delphi Study on key competencies in sustainability (Brundiers et al. 2021). Perhaps, as Wiek and Redman suggest in Chap. 4, it is time to stop creating frameworks and focus instead on their implementation.

Four Hazards of the Competence Approach

The accelerating rate at which global environmental systems are destabilising suggests that wholesale societal change, managed or otherwise, will be inevitable. Education clearly has a critical role to play in any sustainable socioeconomic pathway (Samir and Lutz 2017), both in reorienting our models of development and preparing us for those changes. This is a serious responsibility that behoves us to subject any proposed educational approach to critique; despite, or indeed because of, its ubiquity, the Competence Turn is no exception. The following discussion therefore identifies four inter-related hazards that an emphasis on competences in environmental and sustainability education might present. These are:

- conceptual confusion,
- the centrality of context,
- impacts on pedagogy,
- distractions from fundamental issues.

The term hazard is a deliberate choice, it suggests situations to be aware of as ever-present dangers that, while being unavoidable, may be managed with caution.

Conceptual Confusion

The term competence is explored at length in Chap. 1 while the difficulty of combining, in the same term, the acquisition of specific skills and knowledge *and* the willingness to use them is addressed with some panache by Kerry Shephard in Chap. 6. Suffice to say we cannot assume that the word 'competence' is uniformly understood, or accepted, internationally. This may seem surprising given that competence-based learning (CBL) has become so widespread precisely because of the sense of certainty, in terms of measurable outcomes, that it provides to education managers.

As we explore in Chap. 1, the meaning of competence can be expanded to include values, attitudes, judgements and motivations as well as skills and knowledge but this has led to varying definitions in different locations. While it may not be realistic to expect a globally agreed definition to emerge, any confusion around meaning is likely to hamper efforts to share learning from place to place. More worryingly, this lack of consensus renders inter-researcher agreement difficult if not impossible across international settings, something that is essential if the implementation of CBL is to be verified by robust comparative research. This may not be a barrier to the implementation of CBL in any given context but this does need taking into consideration when evaluating proposals based on experiences of competence-based approaches elsewhere.

To some extent, a heightened awareness of this difficulty can be beneficial. If the advent of post-modernism gave us anything, it is the recognition that the replication of social processes is likely to be impossible; context matters. Acknowledgement that 'competence' is not universally understood should serve to remind policy-makers, educators and researchers alike of the need to unpack their assumptions when proposing competence-based approaches, to clarify exactly what they mean, how they intend to go about it and, crucially, to what ends.

The Centrality of Context

Even if the concept of competence were to be firmly pinned down, it is unlikely that any universally agreed definition would be a comfortable fit with the underpinning principles of environmental and sustainability education (ESE). A competent education system (in the Type 3 sense—see Chap. 1) staffed by competent teachers may be a prerequisite for a sustainable society yet it is unlikely to be the specific combination of competences that is crucial here, rather it is the context in which those competences are applied that will render them 'sustainable'—or not. All ESE-related competence frameworks, for example, include some form of systemic thinking (Corres et al. 2020; Rieckmann 2018), yet there will be many situations, such as the development of specific technologies, that instead require a highly focused systematic approach, which in isolation can be seen as problematic (Sterling 2001).

Any competence needs to be context bound in order to be put into practice, yet a competence for transforming the current state of affairs, let alone a competence for thriving in a sustainable society that has yet to exist is, by definition, difficult to practice in an authentic setting. Indeed, if an ideal situation for practising these competences already existed, there would not be such an urgent need to teach them. Even familiar competences, such as critical thinking, will develop over time and with practice in different contexts rather than in classroom settings (see Chap. 7).

This poses a double bind for any framework of competences for sustainability educators: the widespread appeal of competences lies in the way in which they define predetermined learning outcomes or goals, yet our best hope for a sustainable future is to educate citizens to be open to unforeseen conditions, to learn our way forward into an unknowable future. This suggests that we would do well to modify our approach to competences. Rather than seeing them as the end goal, a 'curriculum as product', we could adopt a more emergentist approach, something Kelly (2009) terms *curriculum as process*. This avoids viewing teachers as technicians delivering predetermined outcomes and instead positions them as facilitators of knowledge production and values identification. It is this process, together with its preparation, teaching and assessment that defines the resultant curriculum. Such a view is at odds with carefully defined competences; it is also difficult to imagine this being condoned within any education system whose overriding purpose is to provide credentials for labour market entry.

A middle way might be to view a competence framework as a proposal to be explored and challenged. This echoes the view of Lawrence Stenhouse (1975) who saw the curriculum, not as a body of material to be covered but as:

> ...a way of translating any educational idea into a hypothesis testable in practice. It invites critical testing rather than acceptance (*Ibid*, p. 142)

This avoids slavish adherence to any given framework. Indeed, such frameworks are most useful as comparators to be reflected on in light of the teaching context, which includes the state of knowledge and ideas of the learners themselves.

One approach to achieving this level of flexibility is that taken by A Rounder Sense of Purpose (RSP—see Chap. 5), which presents its competences in the form of an artist's palette.[1] This invites creativity on the part of the educator, encouraging them to combine competences in unique, context-specific ways while allowing space for additional competences to be added as new ideas are developed. In this way the palette avoids the sense of a linear progression in favour of an emergent approach, accepting that each time the competences are used they will support a unique learning episode with its own unforeseen outcomes.

Impacts on Pedagogy

A number of different pedagogical strategies are discussed in Part III of this volume beginning with an overview by Lozano and Barreiro-Gen (Chap. 17) whose work suggests that the available range of approaches is under-used or not applied appropriately (see Chap. 20). It seems ironic that sustainability competences, which are generally framed as observable behaviours, should be taught predominantly through transmissive approaches such as lectures rather than through more practical means. Perhaps this should not be surprising given that this research focuses on higher education. In any formal education setting however, the way that competences are often broken down into knowledge, skills, judgements, values and so forth leads to detailed inventories that can atomise learning in a manner that is antithetical to the holistic ethos of sustainability.

Equally concerning is the impact that this outcomes-based approach can have on the learner's view of their role in the world. Even as students enumerate the competences that they are

[1] https://aroundersenseofpurpose.eu/framework/palette.

acquiring, a slavish adherence to such carefully prescribed learning outcomes offers little hope for those who might wish to challenge the *status quo*. This is something that Biesta (2015) identifies as a widespread tendency across formal educational systems; by inhibiting students from challenging accepted norms, education suppresses the emergence of human uniqueness. Curriculum as product approaches certainly present this danger, making it more difficult for teachers to facilitate learner agency, a key principle of ESE. As Dewey (1916) reminds us, objective-based learning can have serious unintended consequences:

> Aims limit intelligence (because) given ready-made, they must be imposed by some authority external to intelligence, leaving to the latter nothing but a mechanical choice of means. (*Ibid*, p. 138)

Instead, Dewey suggests the use of objectives as a heuristic device, what Dewey terms a 'legitimate aim'. As discussed above in relation to using frameworks in context, this reflects an emergentist approach:

> The value of a legitimate aim… lies in the fact that we can use it to change conditions. It is a method for dealing with conditions so as to effect desirable alterations in them. (Dewey 1916, p. 138)

This pragmatist approach underpins a model suggested by Öhman and Sund (2021) that avoids the tendency to aggregate components of learning into competences by disaggregating the concept back into its constituent parts. In this case they use the concept of 'sustainability commitment' proposed by Scott (2011), which they see as comprising three inter-related aspects: intellectual, emotional and practical.

> The model suggests that sustainability commitment should be a common goal for ESE and that a sound commitment is situated at the intersection of the intellectual, emotional, and practical aspects of sustainability. (Öhman and Sund 2021, p. 16)

They start by considering the pedagogic approaches that might build this layered commitment rather than the defining competences in detail which carry the drawbacks discussed above. Rather than a return to curriculum as content, they propose an iterative process that draws on students' knowledge, thoughts and experiences as well as careful preparation on the part of the teacher.

Distractions from Fundamental Issues

The hazards discussed above are concerned with the way in which the language—and nature—of competences might lead us to adopt approaches to education that run counter to the principles of ESE. Stepping back from the classroom setting, we might ask ourselves what we are *not* including in these discussions. There is a danger that all this focus on competences diverts our attention away from the overarching issue of the purpose of education itself. The framework proposed by A Rounder Sense of Purpose (Vare et al. 2019; Chap. 5) uses the language of competences in order to engage with current debate in this area but the project's name belies a deliberate attempt to raise a broader concern, that is, the need to reframe the purpose of education beyond its narrow, predominantly economic focus.

There are critical and enduring concepts and principles within ESE that might be discussed under the knowledge component of specific competences but which fall outside of the remit of existing competence frameworks. These include considerations of deep ecology (Devall and Sessions 1985), the need to consider the more than human world, perhaps through notions of 'inclusive pluralism' (Kopnina and Cherniak 2016), the possibility of adopting an eco-justice pedagogy (Bowers 2002) or, as Bonnett (2002) has suggested, a complete adjustment to our 'frame of mind' in relation to human-nature understandings. Surely these issues lie at the heart of our current global predicament vis-à-vis the environment, they certainly challenge our current unsustainable model of development. Yet if these issues are raised in a mainstream education discussion, they still appear to be of minority interest, something of concern to the 'green lobby'. Discussion of competences has brought us close to current policy debates, such as those

European competence frameworks listed in the introduction to this chapter. The danger is that we are held at a distance while we focus on which and whether competences are the best way forward.

Identifying key ESE competences for educators will, we hope, prove over the long term to have made a positive contribution to our long-term survival; if we did not recognise this possibility, we would not have written this book. However, if there is one competence that any self-respecting educator for sustainability might usefully burnish, it is the ability to maintain a critical eye in the face of any framework that comes their way, even (or especially) those that come with the full weight of official compulsion. If education is about anything, it is surely about opening our minds to the myriad possibilities presented by our changing, uncertain world—and recognising the dangers of focusing on too few of them.

References

Bacigalupo M, Kampylis P, Punie Y and Van Den Brande L. (2016) *EntreComp: The Entrepreneurship Competence Framework*. EUR 27939 EN. Luxembourg (Luxembourg): Publications Office of the European Union. JRC101581 https://doi.org/10.2791/16081110.2791/59388410.2791/29479

Barth, M., Godemann, J., Rieckmann, M. & Stoltenberg, U. (2007). Developing Key Competencies for Sustainable Development in Higher Education. *International Journal of Sustainability in Higher Education* 8 (4), 416–430. https://doi.org/10.1108/14676370710823582

Berry, C. (2021) The Case for Hope. *The New Economics Zine: People. Planet. Power*. Issue 3; 28-30. Accessed July 2021: https://neweconomics.org/uploads/files/NEFZINE-issue3-WEB.pdf

Bertschy, F., Künzli, C. & Lehmann, M. (2013). Teachers' Competencies for the Implementation of Educational Offers in the Field of Education for Sustainable Development. *Sustainability*, 5(12), 5067–5080. https://doi.org/10.3390/su5125067

Biesta, G. (2015). What is education for? On Good Education, Teacher Judgement Educational Professionalism, *European Journal of Education*, 50 (1), 75-87

Bonnett, M. (2002) Education for Sustainability as a Frame of Mind, *Environmental Education Research*, 8:1, 9-20

Bowers, C. A. (2002) Toward an Eco-justice Pedagogy, *Environmental Education Research*, 8:1, 21-34

Brundiers, K., Barth, M., Cebrián, G., Cohen, M., Diaz, L., Doucette-Remington, S., Dripps, W., Habron, G., Harré, N., Jarchow, M., Losch, K., Michel, J., Mochizuki, Y., Rieckmann, M., Parnell, R., Walker, P. & Zint, M. (2021). Key competencies in sustainability in higher education—toward an agreed-upon reference framework. *Sustain Sci* 16, 13–29. https://doi.org/10.1007/s11625-020-00838-2

Corres, A., Rieckmann, M., Espasa, A. & Ruiz-Mallén, I. (2020). Educator Competences in Sustainability Education: A Systematic Review of Frameworks. *Sustainability*, *12*(23), 9858. https://doi.org/10.3390/su12239858

de Haan, G. (2006). The BLK '21' programme in Germany: a 'Gestaltungskompetenz'-based model for Education for Sustainable Development, *Environmental Education Research*, 12:1, 19-32, https://doi.org/10.1080/13504620500526362

Devall, B. & Sessions, G. (1985) *Deep Ecology: living as if nature mattered*. Salt Lake City: Peregrine Smith Books.

Dewey J (1916) *Democracy and Education*. Project Gutenberg EBook. Accessed at: www.gutenburg.org

Harvey, D. (2005) *A Brief History of Neoliberalism*. Oxford: OUP

Kelly, A. V. (2009) *The Curriculum Theory and Practice* (6th edition). London: Sage

Klieme, E., Hartig, J. & Rauch, D. (2008). "The concept of competence in educational contexts". In J. Hartig, E. Klieme and D. Leutner (eds.) *Assessment of competencies in educational settings. State of the art and future prospects*. Cambridge, UK: Cambridge University Press, pp. 3-22.

Kopnina, H. & Cherniak, B. (2016) Neoliberalism and justice in education for sustainable development: a call for inclusive pluralism. *Environmental Education Research*, 22: 6, 827852, https://doi.org/10.1080/13504622.2016.1149550

Lawton, D. (1996) *Beyond the National Curriculum: teacher professionalism and empowerment*. London: Hodder and Stoughton

OECD (2002). *Definition and Selection of Competencies (DeSeCo): theoretical and conceptual foundations*. Directorate for Education, Employment, Labour and Social Affairs, Education Committee, DEELSA/ED/CERI/CD (2002) 9, 27. pp.

Öhman, J. & Sund, L. A (2021) Didactic Model of Sustainability Commitment. *Sustainability*, 13, 3083. https://doi.org/10.3390/su13063083

Punie, Y. & Brecko, B., editor(s), Ferrari, A., (2013) *DIGCOMP: A Framework for Developing and Understanding Digital Competence in Europe*. EUR 26035, Publications Office of the European Union, Luxembourg, ISBN 978-92-79-31465-0, doi:https://doi.org/10.2788/52966, JRC83167

Rauch, F. & Steiner, R. (2013). Competences for education for sustainable development in teacher education. *CEPS Journal*, 3, 9–24

Rieckmann, M. (2018) Learning to transform the world: Key competences in Education for Sustainable Development. In *Issues and Trends in Education for Sustainable Development*; Leicht, A., Heiss, J., Byun,W.J., Eds.; UNESCO: Paris.

Rychen, D. S. (2019) *Alignment with OECD Definition and Selection of Competencies: Theoretical and Conceptual Foundations* (DeSeCo) Project. Paris: OECD. https://www.oecd.org/education/2030-project/teaching-and-learning/learning/transformative-competencies/Thought_leader_written_statement_Rychen.pdf

Sala, A., Punie, Y., Garkov, V. & Cabrera Giraldez, M. (2020) *LifeComp: The European Framework for Personal, Social and Learning to Learn Key Competence*, EUR 30246 EN, Publications Office of the European Union, Luxembourg, ISBN 978-92-76-19417-0, doi:https://doi.org/10.2760/922681, JRC120911.

Samir, K.C. & Lutz, W. (2017). The human core of the Shared Socioeconomic Pathways: population scenarios by age, sex and level of education for all countries to 2100. *Global Environ. Change,* 42,181-192. doi: https://doi.org/10.1016/j.gloenvcha.2014.06.004

Scott, W. A. H. (2011) Sustainable schools and the exercising of responsible citizenship—A review essay. *Environmental Education Research*, 17, 409–423. https://doi.org/10.1080/13504622.2010.535724

Sleurs, W. (Ed.) (2008) *Competencies for ESD (Education for Sustainable Development) Teachers, a Framework to Integrate ESD in the Curriculum of Teacher Training Institutes.* Comenius 2.1 project 118277-c p-1-2004-b e-Comenius-c2.1. http://www.unece.org/env/esd/inf.meeting.docs/egonInd/8mtg/csct%20HandbookExtract.pdf

Stenhouse, L. (1975). *An introduction to Curriculum Research and Development.* London: Heinemann.

Sterling, S. (2001). *Sustainable Education, Re-visioning Learning and Change*. Dartington: Green Books

Timm, J.-M. & Barth, M. (2021). Making education for sustainable development happen in elementary schools: the role of teachers. *Environmental Education Research*, 27(1), 50–66. https://doi.org/10.1080/13504622.2020.1813256

UNECE (2009) *Learning from Each Other: The UNECE Strategy for Education for Sustainable Development,* Geneva: UNECE. https://sustainabledevelopment.un.org/content/documents/798ece5.pdf

UNECE (2012) *Learning for the Future: Competences in Education for Sustainable Development,* Geneva: United Nations. https://unece.org/fileadmin/DAM/env/esd/ESD_Publications/Competences_Publication.pdf

UNESCO-UNEP (1978) *Inter-governmental Conference on Environmental Education*, October 14–26, 1977, Tbilisi. Paris: UNESCO-UNEP.

Vare, P., Arro, G., Hamer, A. de, Del Gobbo, G., Vries, G. de, Farioli, F., Kadji-Beltran, C., Kangur, M., Mayer, M., Millican, R., Nijdam, C., Réti, M. & Zachariou, A. (2019). Devising a Competence-Based Training Program for Educators of Sustainable Development: Lessons Learned. *Sustainability*, *11*(7), 1890. https://doi.org/10.3390/su11071890

Wiek, A.; Withycombe, L.; Redman, C.L. (2011) Key competences in sustainability: A reference framework for academic program development. *Sustainability Science*, 6, 203–218. Key competencies in sustainability: a reference framework for academic program development | SpringerLink

Wilke, R.J., Peyton, R.B. & Hungerford, H.R. (1987) *Strategies for the training of teachers in environmental education.* International Environmental Education Programme; environmental education series No. 25. Paris: UNESCO-UNEP.

3 Educators' Competence Frameworks in Education for Sustainable Development

Marco Rieckmann and Matthias Barth

Abstract

Educators play an important role in the implementation of Education for Sustainable Development (ESD). However, they need to be qualified to work with the concept of ESD, acquiring specific competences to deal with sustainable development issues and to align their pedagogical practice with ESD. This chapter describes different frameworks and models for ESD competences, putting them in relation to each other and discussing them critically. It also situates the RSP framework in the wider context and clarifies the demands made of ESD educators and the competences they need to develop. Heuristic analysis is undertaken to ensure more systematic investigation of the different ESD competence frameworks, on the basis of two descriptors: (1) target group and how specifically a target group is defined and (2) the relation between content knowledge and pedagogical knowledge. Further research is needed, in particular with regard to the theoretical foundations and the operationalization of the competence frameworks.

Keywords

Education for sustainable development · Educator competences · Professional action competence · Critical review

Introduction

Education for Sustainable Development (ESD) promotes the development of sustainability competences (Brundiers et al. 2021; Rieckmann 2018; Lozano et al. 2017; see Chap. 4 in this book) with a view to addressing the manifold ecological, social, economic, and cultural problems in the world and effecting the necessary societal transformation (UN Environment 2019). ESD enables people to participate in sustainable development and to reflect critically on their own actions. This does not mean prescribing particular ways of thinking or behaving, but on the contrary empowering individuals to think about sustainable development issues for themselves and to find their own answers (Rieckmann 2018; Wals 2015).

Educators are powerful agents for change, delivering the educational response required to

M. Rieckmann (✉)
University of Vechta, Vechta, Germany
e-mail: Marco.Rieckmann@uni-vechta.de

M. Barth
Eberswalde University for Sustainable Development, Eberswalde, Germany

P. Vare et al. (eds.), *Competences in Education for Sustainable Development*, Sustainable Development Goals Series, https://doi.org/10.1007/978-3-030-91055-6_3

achieve sustainable development in general and the Sustainable Development Goals (SDGs) in particular (UNESCO 2017). Whether education processes and educational institutions as such can become more sustainable depends on the knowledge, competences, attitudes and values of educators, but also on their interaction with institutional frameworks and curricular structures.

Teacher education and the education and professional development of educators in general can meet this challenge by reorienting themselves towards ESD as emphasized in various international declarations and national policy papers (UNECE 2005; UNESCO 2009). Various examples of the integration of ESD into teacher education have shown how the support of teachers has been a crucial prerequisite for the successful adoption and implementation of ESD (UNESCO 2014).

However, efforts to prepare educators to implement ESD have not advanced sufficiently and more still need to be done to refocus educator support for ESD in terms of content and teaching and learning methods. This is why priority action area 3 of the UNESCO *ESD for 2030* programme aims to build the capacities of educators (UNESCO 2020). This priority action area focuses on fostering the competences needed by change agents to promote ESD, and integrating ESD into the education and training of early childhood, elementary and secondary and vocational education teachers and trainers (UNESCO 2020).

This chapter deals first with competence-based education and the concept of professional competence and then goes on to describe various ESD competence frameworks. It thus locates the Rounder Sense of Purpose[1] (RSP) competence concept in the discourse on ESD competences. A systematic analysis of the different ESD competence frameworks is then undertaken. Finally, the need for further development of the competence concepts and for further research is identified.

Competences in Educational Programmes

Competences include knowledge, skills, motivation, attitudes and values systems and enable individuals to perform tasks successfully and solve problems in different situations and contexts (Vare et al. 2019; Rieckmann 2012). Competence-based education focuses on students' learning outcomes rather than on what teachers should be teaching (see Chap. 2). "For the teacher, competences help them to become able to perform better and more effectively under different circumstances, frameworks and conditions" (Vare et al. 2019, p. 2).

Competences cannot simply be taught; they have to be developed (Weinert 2001). ESD therefore requires a transformative, action-oriented pedagogy (Rieckmann 2018; Barth 2015; see Chap. 17). This entails a twofold challenge for programmes and activities aiming to educate the educator. First, it requires a thorough understanding of the competences learners should be able to develop. Second, it calls for educators to have the competences to support the competence development of the learners and it thus raises the question of what competences are needed by educators (Brandt et al. 2019).

Frameworks that take the role of educators into account largely build on Shulman's (1987) categories of what constitutes a competent teacher, differentiating between content knowledge ("what to teach") and pedagogical content knowledge ("how to teach"). In an empirically tested and widely adopted approach, Baumert and Kunter (2013) designed a model for teachers' professional competence, identifying professional knowledge, beliefs, motivation, and self-regulation as core aspects (Baumert and Kunter 2013).

[1] https://aroundersenseofpurpose.eu/.

Frameworks and Models of Educators' Competences in ESD

The implementation of ESD in any kind of educational institution places high demands on educators and presupposes that they have dealt with the concepts of sustainable development and ESD. Integrating the concept of ESD into pedagogical training empowers educators to address key societal issues and to deal with them together with their students. "Educators in all educational settings can help learners understand the complex choices that sustainable development requires and motivate them to transform themselves and society" (UNESCO 2020, p. 30).

Integration of ESD enables educators to design learning processes to support the acquisition of sustainability competences in the classroom. Qualifying educators to work with the ESD concept can be expected to contribute to innovations in education and an increase in the quality of education (Barth and Rieckmann 2012). This is reiterated by UNECE, which devotes two indicators to teacher education under "equipping educators with the competence to include sustainable development in their teaching": Sub-indicator 3.1.1: Is ESD a part of the initial educators' training? And sub-indicator 3.1.2: Is ESD a part of the educators' in-service training? (UNECE Expert Group 2007, p. 7).

It is widely agreed that educators need to be qualified to work with the concept of ESD and that they should acquire specific competences in order to deal with sustainable development issues and to align their pedagogical practice with this concept. "This includes understanding key aspects of each of the 17 SDGs and their interlinkages, as well as understanding how transformative actions occur and which […] transformative pedagogical approaches can best bring them about" (UNESCO 2020, p. 30). ESD can encourage educators to consider their pedagogical practice from a new perspective. In order for educators to be prepared to deliver ESD, they need to develop key sustainability competences (including knowledge, skills, attitudes, values, motivation and commitment) (Brundiers et al. 2021; Rieckmann 2018; see Chap. 4). However, in addition to general sustainability competences, they also need ESD competences, which can be defined as a capacity to support the development of sustainability competences through a range of innovative teaching and learning practices (Corres et al. 2020; Bertschy et al. 2013). With regard to the development of competences by educators, the German National Action Plan on ESD calls on the Federal State to "promote the development of ESD competence models for teacher education (school and university teachers)" and to "promote pilot projects on teacher education for sustainable development at all phases, and the interrelations between them" (National Platform ESD 2017, p. 27).

In order to integrate ESD into the classroom and to promote sustainability competences among students, educators should be able to address the challenges of sustainable development and examine their own role in this process. Issues raised by the concept should be considered in an integrative and cross-disciplinary way, and from multiple perspectives. Educators should know about sustainable development, the different SDGs and the related topics and challenges, and should reflect on the concept of sustainable development, the challenges of achieving the SGDs, the importance of their own field of expertise for achieving the SDGs and their own role in this process. They should also understand the discourse on and the practice of ESD in its local, national, and global context. In addition, educators should be able to design (formal, non-formal, and informal) learning environments that allow students to participate in and gain experience of sustainable development tasks, taking an action-oriented and transformative approach to teaching. Furthermore, they should act as change agents in a process of organizational learning that advances their educational institution towards sustainable development (UNESCO 2017).

These elements of ESD competence are described in greater detail in a number of different frameworks for educators' ESD competences, such as the CSCT model (Sleurs 2008), the UNECE framework (UNECE 2012), the KOM-

BiNE model (Rauch and Steiner 2013), and the approaches of Bertschy et al. (2013) and Timm and Barth (2021). Teacher education and any other educator training programmes should be further developed to meet these standards.

The *CSCT competence model* focuses on teachers as individuals, as participants in educational institutions and as members of a particular society, i.e. it refers to teachers' personal and social behaviour as well as their professional role. It describes ESD competences through three superordinate dimensions (teaching/communicating, reflecting/visioning, networking) and five competence domains (knowledge, systems thinking, emotions, values and ethics, action) (Sleurs 2008). The shortcomings of this model relate, for example, to the fact that "the distinction between the five domains of competency is rather vague and unconvincing in parts. For example, the domain 'emotions' is problematic as emotions play a role as a concomitant in all the other domains [...]" (Bertschy et al. 2013, p. 5069).

The *UNECE framework* (2012) covers all educational practitioners and includes 39 competences, presented in four domains of learning: 1. Learning to know (The educator understands...), 2. Learning to do (The educator is able to...), 3. Learning to live together (The educator works with others in ways that...), 4. Learning to be (The educator is someone who...); and with three principles: holistic approach, envisioning change, and achieving transformation. "The UNECE framework represents a significant attempt to identify competences in a systematic and comprehensive manner with the explicit aim of becoming a commonly shared reference framework" (Vare et al. 2019, p. 6).

The *KOM-BiNE model* (Rauch and Steiner 2013) "is not based on individuals, but on a group whose members pool their competencies for ESD in specific projects or issues and act as a team" (p. 16). It includes the following competence fields: knowing and acting, valuing and feeling, communicating and reflecting, visioning, planning and organizing, and networking. It refers to three different fields of action: instruction, participation in the design of the educational institution, and reaching out to society, to the institution's closer and wider environment.

Bertschy et al. (2013) present a competence model for *ESD-specific professional action competences for teachers in Kindergarten and Primary School.* Based on Baumert and Kunter's (2013) model of "professional action competence for teachers", this ESD competence model describes four aspects of competence (professional knowledge, motivation, conviction/values, self-regulation) and five fields of competence (pedagogical knowledge, content knowledge, pedagogical content knowledge, organizational knowledge, guidance knowledge).

Timm and Barth (2021) recently supplemented the discussion on theoretically based competence models with an empirical analysis giving a voice to teachers who are experienced in ESD. On the basis of interviews with teachers involved in ESD at German elementary schools, they identified *two types of ESD teachers and their respective competence profiles*: teachers who function as change agents by interacting with students (in-class teachers), and teachers who function as change agents by inciting institutional change (structure-focused teachers). The authors found significant differences between the two groups with regard to their level of activities, their perspective on teaching and their understanding of content knowledge and pedagogical content knowledge. Bearing these different types of change agents in mind provides a more nuanced picture of potential competence profiles and offers important pointers for educational programmes (Timm & Barth 2021).

Assuming that "the UNECE competences were abstract, complex, and repetitious, and that the total of 39 was simply unmanageable" (Vare et al. 2019, p. 6), the *European project 'A Rounder Sense of Purpose' (RSP)* "set out to develop and test a framework of educator competences that could contribute to achieving a more sustainable world" (Vare et al. 2019, p. 2; see Chap. 5). This is a conceptual framework that has been designed for all educators, working at any level, who wish to provide ESD. Educators need knowledge of innovative teaching and learning methods, but also need the competences to apply them. In addition, they need competences to support students, for example, with projects, which also implies reflecting critically on their own role as educators and seeing themselves more as facilitators of learning. Educators need to adopt a critical stance and be able to assess and evaluate students' development in this area. It is also important to recognize that under the RSP framework ESD competences are seen as mutually supportive and therefore not in isolation. The RSP framework consists of twelve ESD competences across three domains (holistic approach, envisioning change, and achieving transformation) and four phases (integration, involvement, practice, and reflection) (Chap. 5). It can be used as the basis of an educational programme and/or to assess educators who wish to improve their ability to contribute to ESD.

Discussion

A recent systematic literature review of ESD competence frameworks and models (Corres et al. 2020) shows that most are based on or related to the UNECE framework and that competences such as Critical Thinking ("Critical contextualization of knowledge establishing interrelationships between social, economic and environmental, local and/or global problems", Rauch and Steiner 2013, p. 19), Participation in Community ("Participation in community processes that promote sustainability", Carracedo et al. 2018, p. 6), and Connections ("To know the main concepts and principles in connection with the Earth as a biophysical system and in connection with the relationships and interactions between society and the environment", Álvarez-García et al. 2019, p. 4) are included in most of the frameworks. By contrast, competences such as Emotions Management ("To manage emotions and concerns: promoting reflection on one's own emotions as a means to reach a deeper understanding of problems and situations", Cebrián and Junyent 2015, p. 2771), Futures ("It offers ways of addressing and helping to shape the future [...]. It enables individuals to recognize relations and possible evolutions between past, present, and future and envision possible or thinkable futures alternatives and their impact", Vare et al. 2019, p. 10), and Achieving Transformation ("Related to transformation approaches in education, pedagogy and for educators and education systems in all the levels (Learning to know, Learning to live together, Learning to be, Learning to do)", Meyer et al. 2017, p. 740) are included less frequently in the frameworks, which means that the transformative potential of ESD has not been fully realized. In addition, the review concludes that some of the frameworks and models lack clear theoretical foundation, for example, an explicit definition of the concepts of sustainability and competences.

Taking into account the frameworks described in this chapter, the question remains as to how these different approaches can be compared more systematically in order to analyse and discuss similarities and differences. One way we consider promising is to differentiate these approaches against the two descriptors of (1) target group and how specifically a target group is defined and (2) the relation between content knowledge (CK) and pedagogical content knowledge (PCK). Using such a distinction enables an area to be mapped along two axes, in which different approaches can be placed (see Fig. 3.1).

In such an area, then, we can see, for example, that while the approaches developed by Bertschy et al. (2013) and similarly Timm and Barth (2021) clearly focus on specific teachers (kindergarten and primary education in the former, primary education in the latter), other frameworks take a more general approach, focusing on teacher education in general (KOM-BiNE model) or even going beyond teacher education, addressing educators in general (RSP framework).

Similarly, a distinction can be made with regard to general orientation. Here, we see frameworks such as the CSCT model, the UNECE framework and the KOM-BiNE model, which refer to more general aspects of sustainability and sustainability competences, at one end of the spectrum. These frameworks flesh out the relationship of teachers to society, their sustainability values, attitudes and behaviours,

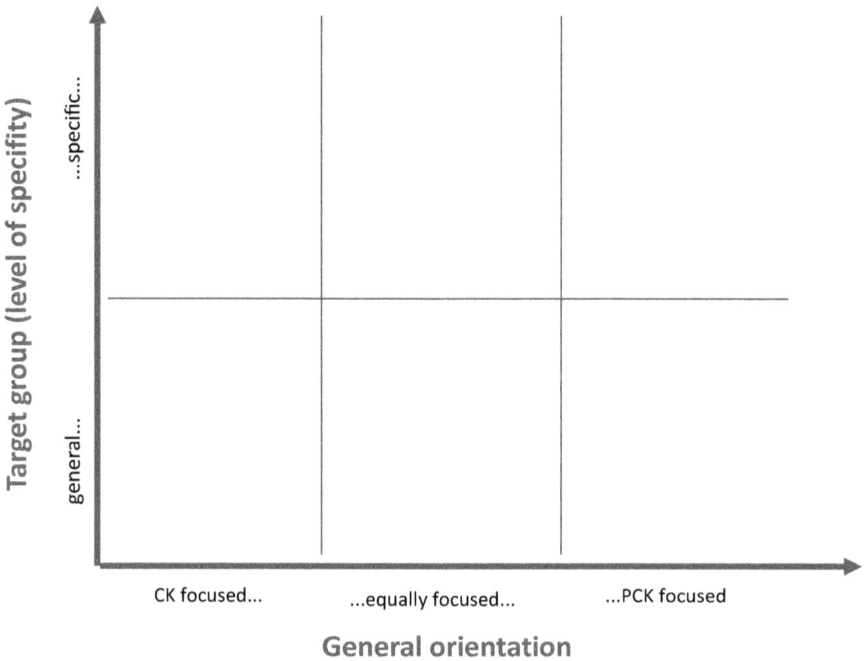

Fig. 3.1 Heuristics for the analysis of ESD competence frameworks and models

and thus their involvement in the sustainable development of society. At the other end of the spectrum is, for example, the approach of Bertschy et al. (2013), which focuses predominately on the professional context and asks what teachers can and need to know in order to be able to teach ESD. Here, however, the question may arise as to whether educators who do not develop sustainability competences themselves and relate, at least to some extent, their own attitudes and values (and behaviours) to sustainable development, are able to work with students on sustainable development issues in a credible way (see Chap. 6). The findings of Timm and Barth (2021) represent a middle ground here, taking both CK and PCK into account. This approach is also reflected in the RSP framework.

Analytical distinction of different frameworks and approaches enables them to be further compared against some key characteristics that come with consequences for educational praxis. We see three main characteristics as being of special interest here:

- the underlying understanding of what being a change agent means,
- the consequences for the design and implementation of educational programmes,
- measurability versus accountability.

The underlying *change agent model* refers to how the frameworks conceive of the role of change agents. This differs significantly between approaches. At one end of the spectrum are frameworks that understand a change agent as someone who can influence education through a different approach to teaching and thus has a narrower understanding of the role of educators as change agents for student empowerment. At the other end of the spectrum are frameworks that take a more holistic approach and see educators as change agents if they teach differently, engage differently in their institutions, and are in general role models for sustainability.

Differences are also evident between the frameworks when it comes to their stance towards educational praxis and thus the *consequences for the design and implementation of educational programmes*. While some frameworks clearly set out to achieve such consequences and come with practical advice on how to implement what, others can be understood as more general frameworks with a more basic focus on objectives but little direct implications for educational praxis.

A final crucial distinction, then, can be drawn between *measurability and accountability*. We see most of the frameworks as general approaches that lack operationalization and do not or do not readily allow for assessment of the development of such competences. This distinction refers rather to accountability for what should be developed and pays less attention to the question of if and how such development can be justified, i.e. measured. The underlying assumption here might be that not all that counts can be measured—an understanding of competence development as emancipatory education we share. It comes, however, with limitations that need to be clearly communicated. On the other side are initial approaches providing examples that can be operationalized and used as assessment procedures to give feedback to both educators and learners. While such an approach again comes with limitations, we see it as an important addition that enables evidence-based design in educational formats to be justified.

Conclusion

Educators play an important role in the implementation of ESD. The ESD competence frameworks and approaches outlined in this chapter provide guidance on the competences that educators should have in order to meet this requirement. Teacher education and the training of educators in general should be more structured around these frameworks, but it should be borne in mind that the different frameworks are each focused on specific target groups. In addition, the relevant programme managers need to take a stand on the question of how far programmes should focus only on the development of professional competences and how far they should also focus on the development of general sustainability competences. Depending on the answer to this question, different frameworks will be more suitable. Both the approach of Timm and Barth (2021) and the RSP framework take an intermediary position here.

Further research is needed, particularly with regard to theoretical foundations and the operationalization of the competence frameworks. The latter aspect is central to ensuring the measurability of ESD competence development. In addition, a research gap can also be identified with regard to the empirical investigation of the relationship between educator competences and student performance.

References

Álvarez-García, O., García-Escudero, L.Á., Salvà-Mut, F. & Calvo-Sastre, A. (2019). Variables influencing pre-service teacher training in education for sustainable development: A case study of two Spanish universities. *Sustainability*, *11*(16), 4412. https://doi.org/10.3390/su11164412

Barth, M. (2015). Implementing sustainability in higher education. Learning in an age of transformation. London: Routledge.

Barth, M. & Rieckmann, M. (2012). Academic staff development as a catalyst for curriculum change towards education for sustainable development: an output perspective. *Journal of Cleaner Production*, *26*, 28–36. https://doi.org/10.1016/j.jclepro.2011.12.011

Baumert, J. & Kunter, M. (2013). The COACTIV Model of Teachers' Professional Competence. In Kunter, M., Baumert, J., Blum, W., Klusmann, U., Krauss, S. & Neubrand, M. (Eds.), *Mathematics teacher education. v. 8: Cognitive activation in the mathematics classroom and professional competence of teachers* (pp- 28–48). New York, London: Springer.

Bertschy, F., Künzli, C. & Lehmann, M. (2013). Teachers' Competencies for the Implementation of Educational Offers in the Field of Education for Sustainable Development. *Sustainability*, *5*(12), 5067–5080. https://doi.org/10.3390/su5125067

Brandt, J.-O., Bürgener, L. & Redman, Aaron, Barth, Matthias (2019). Becoming a competent teacher in education for sustainable development – learning outcomes and processes in teacher education. *International Journal of Sustainability in Higher Education*, *20*(4), 630–653, https://doi.org/10.1108/IJSHE-10-2018-0183

Brundiers, K., Barth, M., Cebrián, G., Cohen, M., Diaz, L., Doucette-Remington, S., Dripps, W., Habron, G., Harré, N., Jarchow, M., Losch, K., Michel, J., Mochizuki, Y., Rieckmann, M., Parnell, R., Walker, P. & Zint, M. (2021). Key competencies in sustainability in higher education—toward an agreed-upon reference framework. *Sustainability Science*, *16*(1), 13–29. https://doi.org/10.1007/s11625-020-00838-2

Carracedo, F.S., Segalàs, J., Vidal, E., Martin, C., Climent, J., López, D. & Cabré, J. (2018). Improving engineering educators' sustainability competencies by using competency maps. The EDINSOST project. *International Journal of Engineering Education*, *34*(5), 1527–1537.

Cebrián, G. & Junyent, M. (2015). Competencies in education for sustainable development: Exploring the student teachers' views. *Sustainability*, *7*(3), 2768–2786. https://doi.org/10.3390/su7032768

Corres, A., Rieckmann, M., Espasa, A. & Ruiz-Mallén, I. (2020). Educator Competences in Sustainability Education: A Systematic Review of Frameworks. *Sustainability*, *12*(23), 9858. https://doi.org/10.3390/su12239858

Lozano, R., Merrill, M., Sammalisto, K., Ceulemans, K. & Lozano, F. (2017). Connecting Competences and Pedagogical Approaches for Sustainable Development in Higher Education: A Literature Review and Framework Proposal. *Sustainability*, *9*(11), 1889. https://doi.org/10.3390/su9101889

Meyer, J., Mader, M., Zimmermann, F. & Çabiri, K. (2017). Training sessions fostering transdisciplinary collaboration for sustainable development: Albania and Kosovo case studies. International Journal of Sustainability in Higher Education, 18(5), 738–757. https://doi.org/10.1108/IJSHE-02-2016-0032

National Platform ESD (Ed.) (2017). *Nationaler Aktionsplan Bildung für nachhaltige Entwicklung.* BMBF. http://www.bne-por-tal.de/sites/default/files/downloads/publikationen/Nationaler_Aktionsplan_Bildung_fuer_nachhaltige_Entwicklung_0817.pdf

Rauch, F. & Steiner, R. (2013). Competences for education for sustainable development in teacher education. *CEPS Journal*, *3*, 9–24

Rieckmann, M. (2012). Future-oriented higher education: Which key competencies should be fostered through university teaching and learning? *Futures*, *44*, 127–135

Rieckmann, M. (2018). Chapter 2 - Learning to transform the world: key competencies in ESD. In A. Leicht, J. Heiss & W. J. Byun (Eds.), *Education on the move. Issues and trends in education for sustainable development* (pp. 39–59). United Nations Educational, Scientific and Cultural Organization

Shulman, L. (1987). Knowledge and Teaching: Foundations of the New Reform. *Harvard Educational Review 57*, 1, 1–23.

Sleurs, W. (2008). *Competencies for ESD (Education for Sustainable Development) teachers. A framework to integrate ESD in the curriculum of teacher training institutes.* Comenius 2.1 project; Katholieke Hogeschool Leuve. http://www.unece.org/fileadmin/DAM/env/esd/inf.meeting.docs/EGonInd/8mtg/CSCT%20Handbook_Extract.pdf

Timm, J.-M. & Barth, M. (2021). Making education for sustainable development happen in elementary schools: the role of teachers. *Environmental Education Research*, *27*(1), 50–66. https://doi.org/10.1080/13504622.2020.1813256

UNECE – United Nations Economic Commission for Europe. (2005). *UNECE Strategy for Education for Sustainable Development.* https://www.unece.org/fileadmin/DAM/env/documents/2005/cep/ac.13/cep.ac.13.2005.3.rev.1.e.pdf

UNECE – United Nations Economic Commission for Europe. (2012). *Learning for the future. Competences in ESD for educators.* http://www.unece.org/fileadmin/DAM/env/esd/ESD_Publications/Competences_Publication.pdf

UNECE Expert Group (2007). *Indicators for Education for Sustainable Development. Progress report on the work of the Expert Group*, http://www.unece.org/fileadmin/DAM/env/documents/2006/ece/cep/ac.13/ece.cep.ac.13.2006.5.e.pdf

UN Environment (Ed.). (2019). *Global Environment Outlook GEO-6: Healthy Planet, Healthy People.* Cambridge University Press. https://www.unenvironment.org/resources/global-environment-outlook-6

UNESCO – United Nations Educational, Scientific and Cultural Organization (2014). *Shaping the Future We Want. UN Decade of Education for Sustainable Development (2005-2014). Final Report.* UNESCO. https://unesdoc.unesco.org/ark:/48223/pf0000230171

UNESCO – United Nations Educational, Scientific and Cultural Organization (2017). *Education for Sustainable Development Goals: learning objectives.* UNESCO. https://unesdoc.unesco.org/ark:/48223/pf0000247444_eng

UNESCO – United Nations Educational, Scientific and Cultural Organization (2020). *Education for sustainable development: a roadmap. ESD for 2030.* UNESCO. https://unesdoc.unesco.org/ark:/48223/pf0000374802

UNESCO World Conference on Education for Sustainable Development. (2009). *Bonn Declaration.* UNESCO. http://www.esd-world-conference-2009.org/fileadmin/download/ESD2009_BonnDeclaration080409.pdf

Vare, P., Arro, G., Hamer, A. de, Del Gobbo, G., Vries, G. de, Farioli, F., Kadji-Beltran, C., Kangur, M., Mayer, M., Millican, R., Nijdam, C., Réti, M. & Zachariou, A. (2019). Devising a Competence-Based Training Program for Educators of Sustainable Development: Lessons Learned. *Sustainability*, *11*(7), 1890. https://doi.org/10.3390/su11071890

Wals, A. E. J. (2015). *Beyond Unreasonable doubt. Education and learning for socio-ecological sustainability in the anthropocene.* Wageningen University. https://arjenwals.files.wordpress.com/2016/02/8412100972_rvb_inauguratie-wals_oratieboekje_v02.pdf

Weinert, F. E. (2001). Concept of Competence: A Conceptual Clarification. In D. S. Rychen & L. H. Salganik (Eds.), *Defining and Selecting Key Competencies* (pp. 45–65). Hogrefe and Huber.

4 What Do Key Competencies in Sustainability Offer and How to Use Them

Arnim Wiek and Aaron Redman

Abstract

Scholars and educators largely agree on a framework that integrates a small set of key competencies in sustainability as learning objectives for courses and programs in higher education. However, the current practice of using these key competencies often falls short due to insufficient competence coverage and integration, competence operationalization, alignment between competencies and pedagogies, assessment of competence development, and assessment of professional success. Herein, we briefly review these shortcomings and outline ways to overcome them, including institutional reforms—providing guidance to scholars, educators, and administrators on how to improve competencies-based educational practice and better contribute to advancing sustainability.

A. Wiek (✉)
School of Sustainability, College of Global Futures, Arizona State University, Tempe, AZ, USA

Center for Global Sustainability and Cultural Transformation, Arizona State University, Tempe, AZ, USA
e-mail: Arnim.Wiek@asu.edu

A. Redman
School of Sustainability, College of Global Futures, Arizona State University, Tempe, AZ, USA

Keywords

Competencies framework · Integration · Assessment · Institutional reform · Higher education

Agreement on Key Competencies in Sustainability

Multiple recent literature and expert studies have confirmed convergence among educational scholars and practitioners on a framework for key competencies in sustainability that can guide academic program development and course design, as well as assessment and accreditation procedures. This framework was first introduced a decade ago by Wiek et al. (2011a) and then further developed in Wiek et al. (2016). It caught the attention of educational scholars and practitioners as indicated in a recent bibliometric study by Grosseck et al. (2019) identifying the Wiek et al. (2011a) article as the "most influential paper" (p. 26) in the field of Education for Sustainable Development (ESD) in the past decade. A recent study by the United States National Academies (2020) used this framework, which is intended to enable students to solve sustainability problems (or, in other words, "to design, implement, and lead proactive change toward a sustainable world," p. 114), as a main reference for developing their recommendations on "strengthening

P. Vare et al. (eds.), *Competences in Education for Sustainable Development*, Sustainable Development Goals Series, https://doi.org/10.1007/978-3-030-91055-6_4

sustainability curricula and programs in higher education [...] and developing a sustainability workforce" (p. viii). A recent study by the European Commission's Joint Research Centre (Bianchi, 2020) refers to this framework as the one that has "set the stage for future works and has often been used by scholars as the foundation for any attempt to describe sustainability competences" (p. 13). Brundiers et al. (2021) conducted a Delphi study with experts from around the world on this framework, revealing broad agreement on "the purpose, namely, to enable and empower students to become effective in positively contributing to sustainability problem-solving [...]; [...] defining the competency-related knowledge, skills, motives, and attitudes independent of and complementary to sustainability topics; the integration of the key competencies into a sustainability problem-solving framework [...]; and the introductory set of learning objectives for each of the key competencies." And a recent review of the peer-reviewed literature 2011–2020 (Redman and Wiek 2021) found that Wiek et al. (2011a) have been cited by 63% (n = 141) of the sampled articles published after it (*n* = 225)—that is far more than any other article—and that it has "facilitated explicit convergence," being used as full foundation in 32 articles (14% of the sampled articles) and as partial foundation in 78 articles (35%).

Wiek et al. (2011a and 2016) synthesized literature into a framework of sustainability problem-solving competence, integrating six key competencies, namely systems thinking, anticipatory, normative, strategic, interpersonal/collaborative, and integrated problem-solving competencies, plus basic competencies such as critical-thinking competence. The framework was operationalized into detailed learning objectives in Wiek et al. (2016). The core of this framework has been consistently confirmed. For example, in a UNESCO publication, Rieckmann (2018) describes the conversion of competencies literature as follows (cf. UNESCO 2017): "There is general agreement within the international ESD discourse that the following key sustainability competencies are of particular importance for thinking and acting in favour of sustainable development: systems thinking competency [...], anticipatory competency [...], normative competency [...], strategic competency [...], collaboration competency [...], critical thinking competency [...], self-awareness competency [...], integrated problem-solving competency [...]" (pp. 44–45). Two recent comprehensive studies convened experts (Brundiers et al. 2021) and reviewed the literature (Redman and Wiek 2021) to consolidate the many proposals for expanding the key competencies framework which have been made over the past decade. These studies independently suggested adding three competencies, namely intrapersonal competence (cf. self-awareness competence), integration competence, and implementation competence; to hierarchically structure the competencies; and to specify learning objectives for students interested in a career as sustainability researcher; among others. Both studies synthesize their findings into extended versions of the key competencies framework; the one by Redman and Wiek (2021) is captured in Fig. 4.1.

While there is abundant literature on competencies in sustainability, there is only little, by comparison, on the strengths and weaknesses of putting them into practice—which is the focus of the present chapter.

Flaws in Current Practices

While the key competencies are widely used in sustainability courses and programs around the world, implementation is often flawed (e.g., Trencher et al. 2018; Redman et al. 2021). We briefly review some of the prominent pitfalls in adopting the key competencies framework in educational practice. The insights are partly extracted from literature, but mostly based on direct observations or informal exchanges with numerous colleagues from different universities that offer sustainability programs.

Insufficient Competence Coverage and Integration Sustainability programs and courses often disaggregate the key competencies framework and treat it as a "grocery list" from which to

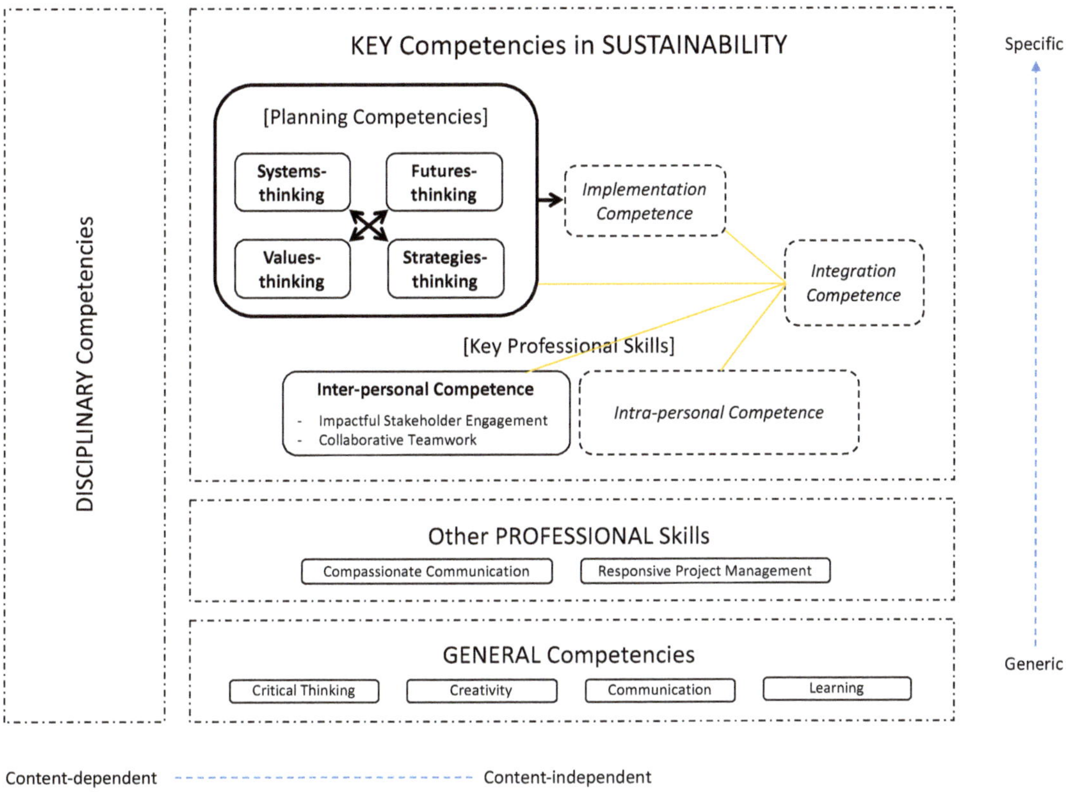

Fig. 4.1 Integrated framework of competencies for advancing sustainability transformations; centered on 8 key competencies in sustainability with 5 established (bold) [based on Wiek et al., 2011b] and 3 emerging (italic); and complemented by disciplinary, general, and other professional competencies (Source: Redman and Wiek 2021)

pick and choose (Redman and Wiek 2021). And scientific reviews have done so, too (e.g., Lozano et al. 2017, p. 4: "Wiek et al. (2011a) compiled a *list* of key competences for SD."). However, a critical feature of this *framework*, as emphasized in all publications on it, is the *integration of* the individual key competencies for sustainability problem-solving. Lack of covering *all* key competencies and conveying how to best *combine* them inevitably leads to students' deficits in sustainability problem-solving competence. Graduates might become proficient systems thinkers, but not sustainability problem-solvers. This uneven coverage of the key competencies was revealed, for example, in a recent study of 45 Master's programs in sustainability, which, for instance, hardly stimulated development of anticipatory competence (Salovaara et al. 2020). In addition, there is a general tendency across sustainability programs and courses to either neglect or underemphasize interpersonal competence development, i.e., teamwork and stakeholder engagement (second least addressed, according to Salovaara et al. 2020). Lip service is being paid to their importance for real-world sustainability problem solving but there are very few sustainability programs that offer a structured pathway for students to develop a proficient level of interpersonal competence concurrently to all other key competencies. The recently suggested additional competencies (intrapersonal competence, etc.) are even less covered and integrated into overall sustainability problem-solving education.

Insufficient Competence Operationalization It continues to be common practice that course instructors and program administrators refer in vague and abstract terms to the key competencies at the beginning of their sustainability course syllabi and course introductions (Brundiers and Wiek 2013). Detailed operationalization of the key competencies (and their inter-play) into specific learning objectives that guide courses throughout the term is often missing. This operationalization is typically left up to instructors who are ill-equipped and supported for the task (Wilhelm et al. 2019). Even if there are program-level learning objectives (PLLOs) determined for sustainability programs, there is rarely a follow-through to what extent these objectives are being covered in the individual courses as well as in the curriculum overall. This leads to a number of deficits: first, instructors and students struggle with distinguishing the key competencies from each other and from other competencies (for example, from general competencies such as critical thinking) as well as clearly relating them to each other; second, there is a gap that remains between the overall learning aspiration (problem-solving competence) and the daily educational practice (isolated competencies); third, students often have a hard time grasping the relevance of the key competencies for their education and future professional practice; and fourth, there is no base for rigorous and comparative assessments of students' acquisition of the key competencies (across different courses and programs).

Insufficient Alignment Between Competencies and Pedagogies As key competencies and learning objectives pertaining to them remain vague, it is almost impossible to adopt the most effective pedagogies to support students in developing them (cf. Lozano et al. 2017). Constructive alignment of pedagogies to match the ambitious and different nature of key competencies (compared to traditional content-based learning objectives) is all too often insufficient (Wilhelm et al. 2019). For example, students' development of anticipatory competence should be supported by imagination and creativity didactics, while normative competencies might best be developed through discursive and deliberative didactics, and interpersonal competencies through experiential and project-based didactics. Additional effort needs to be devoted to pedagogies that build the attitudinal component of key competencies as well as intrapersonal competence. While part of this shortcoming can be credited to the previous one (lack of competence operationalization), many sustainability programs struggle with familiarizing their teaching staff with advanced and up-to-date pedagogical concepts and techniques through advanced trainings as well as supporting and fully embracing their use and implications (Wilhelm et al. 2019). The mantra of "a good researcher is a good teacher" prevails in academia despite plenty of evidence to the contrary. Even if the key competencies are adequately operationalized, insufficient alignment of pedagogies leads to students' deficits in sustainability problem-solving competence. Pointing to the contrary, initial screening and evaluations indicate that there are still significant gaps in supporting students in developing key competencies in sustainability programs (Trencher et al. 2018; Salovaara et al. 2020; Redman et al. 2021).

Insufficient Assessment of Competence Development Many sustainability programs and courses lack rigorous assessments of students' key competencies in sustainability, as, for the most part, scaled self-assessment by the students themselves prevails (Redman et al. 2021). Again, this is due, in part, to deficits mentioned above—an "error reproduction" in constructive alignment of competencies, pedagogies, and assessments. Many instructors retreat to subjective assessment measures (by the students themselves) because the learning objectives are insufficiently operationalized, and the course pedagogy is insufficiently aligned with the learning objectives (cf. Lozano et al. 2017). The current lack of objective assessments results also leads to frustration among students left with few ways to demonstrate success in acquiring the key competencies and little opportunity for making a compelling case to future employers. Current educational practice does not show any regular use of assessment instruments that yield objective results of students' competence acquisition across courses within a program or across programs (to indicate pedagogical quality of courses and programs).

Insufficient Assessment of Professional Success Finally, while there is a significant gap in rigorously assessing students' development of key competencies in sustainability programs, there is practically none in professional practice (Sandri et al. 2018). So, how do we actually know that students have acquired sustainability problem-solving competence through their education? We don't. Success in professional practice through real positive impacts is supposed to be the main reference of successful education—namely, that graduates are actually able to contribute to sustainability problem solving in the real world (not only demonstrate it through proxies in academic programs). Little is being done to evaluate the success of sustainability problem-solvers after graduation, in particular, around the key competencies framework. Obviously, there are logistical challenges, but if we are not able to provide this kind of evidence, doubts will rightfully remain regarding pedagogical effectiveness and efficiency of sustainability courses and programs.

Good Practices

While there are many challenges to effectively applying the key competencies framework in educational practice, there are also a number of robust practices to counter them, which we present below.

Covering and Integrating Competencies Sustainability curricula ought to cover all competencies sufficiently and integrate them so as to deliver on the promise to educate students in sustainability problem-solving competence. The first step is to make sure that all the individual courses do so in conjunction (competencies-oriented curriculum planning). In a second step, some individual courses will need to be designed in a way that they explicitly integrate key competencies in ways which allow students to develop and apply them *in combination* to solve sustainability problems (e.g., in project-based, solution-oriented courses). Advanced (graduate) courses might even cover a number of different sustainability problem-solving frameworks (Wiek and Lang 2016). On the curriculum level, it might be helpful for students and instructors to develop several distinct student "roadmaps" that outline *how* students are expected to successively acquire sustainability problem-solving competence as they move through the program. Considering the importance of interpersonal competence for sustainability professionals, attention should be paid to explicating pathways to develop interpersonal competence over the course of the program (Brundiers et al. 2010).

Operationalizing the Competencies Based on competencies-oriented curriculum planning (see above), it is critical to operationalize the key competencies in specific learning objectives relevant to each course (Brundiers and Wiek 2013). It might help to start from previous operationalizations (Wiek et al. 2016; Brundiers et al. 2021) and adapt them to the specific course content. Course-specific competence operationalization through learning objectives allows for making the key competencies tangible and relevant to students as well as external stakeholders (e.g., auditors, employers). This also enables instructors to conduct rigorous assessments of students' competence acquisition throughout and at the end of the term (Redman et al. 2021). Ideally, the specific learning objectives of each course would be coordinated across the curriculum to secure a sufficient, but not too high level of redundancy (reinforcement).

Aligning Pedagogies to Competencies A success factor for formulating and then teaching toward specific learning objectives directly tied to the key competencies is, obviously, that alignment is sufficiently operationalized

(Lozano et al. 2017) and that the teaching staff is sufficiently trained to implement such alignment (Wiek et al. 2011b; Barth and Rieckmann 2012; Rieckmann 2018). This requires awareness by course instructors that a "one-size-fits-all" approach to pedagogy is inappropriate for teaching the diverse set of key competencies. It also requires from administrators and institutions to put more emphasis on, and develop regular offerings in, initial and continuous pedagogical education for instructors in sustainability programs (Barth and Rieckmann 2012; Scherak and Rieckmann 2020; Weiss et al. 2021). Finally, auditing and evaluations about the suitability of employed pedagogies during courses and ex-post allow for continuous capacity building and improvements of educational practice.

Assessing Competence Development Assessments of students' acquisition of the individual key competencies as well as overall sustainability problem-solving competence can utilize a number of different tools (Rieckmann 2018; Redman et al. 2021). Tools range from *self-perceiving-based* procedures (e.g., reflective writing) that allow students to assess their own competence level and/or development; through *observation-based* procedures (e.g., performance observation) for which instructors or experts assess students' competencies; to *test-based* assessment procedures (e.g., scenario/case test) which use a predefined set of criteria (or "correct" answers) to assess students' competencies. Good practices include making sure that assessments rely on sufficiently operationalized competencies (see above); that instructors are comfortable with the assessment tools through previous familiarization and training; that assessment tools are used in support of students' learning, thus, applied formatively rather than ex-post; and that, ideally, multiple tools are being used in order to triangulate assessment results for increased validity and reliability (particularly relevant when using self-perceiving-based procedures).

Assessing Professional Success Simulating professional situations and asking students to perform within them has been introduced as a pedagogical approach—at times explicitly borrowing from medical, social work, or management education—for advancing students' learning of competencies, including those related to sustainability (Foucrier and Wiek 2020; Redman et al. 2021). This has been called for previously and concurrently to the development of the key competencies concept (Wiek et al. 2011b). It requires significant preparation on the instructors' part, but initial evaluation indicates the added value of these investments (Foucrier and Wiek 2020). Ideally, such pedagogical and assessment approaches would be extended to actual professional practice, for instance, in collaboration with program alumni working as sustainability professionals. A key success factor for this advanced educational practice would be to feed the results back into course and curriculum design for enhanced pedagogical effectiveness and efficiency.

Outlook

Failure or success in adopting the framework of key competencies in sustainability often comes down to the level of institutional support and incentives at universities (Weiss et al. 2021) as well as the wider institutional and political context (United States National Academies 2020). Already a decade ago it was noted and outlined, concurrently with the original development of the key competencies framework, what institutional support might be critical for advancing education for sustainability problem solving (Wiek et al. 2011b). There are a number of supportive institutional structures that universities can adopt, including faculty training, promotion and tenure criteria, and financial incentives (Rieckmann 2018). Institutional inertia and individuals' reluctance have resulted in too few changes in educational practice over the past decade but junior faculty seem to be more willing to adopt good practices, in particular when they were themselves educated in genuine sustainability programs.

While more momentum needs to be created in changing the educational practice and underlying drivers, from institutional support to individual responsibility, one aspect we feel compelled to advise against: *no more reinventing competencies in sustainability!* There is so much work to be done to make the practice of sustainability education more effective and efficient, before running out of time, that all of our collective effort should shift there. The existing convergence on a framework of key competencies in sustainability problem solving seems sufficient for moving forward on advancing the educational practice that the well-being of people and planet depends upon, at least to a significant extent.

Acknowledgements The authors gratefully acknowledge funding from the Lower Saxony Ministry of Science and Culture and the Volkswagen Foundation for the grant "Educating Future Change Agents—Higher Education as a Motor of the Sustainability Transformation" (A115235) through the program "Science for Sustainable Development."

References

Barth, M., & Rieckmann, M. (2012). Academic staff development as a catalyst for curriculum change towards education for sustainable development – an output perspective. *Journal of Cleaner Production*, vol. 26, pp. 28-36.

Brundiers, K., Wiek, A., & Redman, C.L. (2010). Real-world learning opportunities in sustainability – From classroom into the real world. *International Journal of Sustainability in Higher Education*, vol. 11, no. 4, pp. 308-324.

Brundiers, K., & Wiek, A. (2013). Do we teach what we preach? An international comparison of problem- and project-based learning courses in sustainability. *Sustainability*, vol. 5, no. 4, pp. 1725-1746.

Brundiers, K., Barth, M., Cebrián, G., Cohen, M., Diaz, L., et al. (2021). Key competencies in sustainability in higher education – toward an agreed-upon reference framework. *Sustainability Science*, vol. 16, pp. 13-29.

Bianchi, G. (2020). Sustainability Competences – A Systematic Literature Review. European Commission's Joint Research Centre, Luxembourg.

Foucrier, T., & Wiek, A. (2020, in review). Assessing students' competencies in sustainability entrepreneurship through in-vivo simulated professional situations. *Journal of Management Education*.

Grosseck, G., Ţîru, L.G. & Bran, R.A. (2019). Education for Sustainable Development – evolution and perspectives: a bibliometric review of research, 1992–2018. *Sustainability*, vol. 11, 6136.

Lozano, R., Merrill, M.Y., Sammalisto, K., Ceulemans, K., & Lozano, F.J. (2017). Connecting competences and pedagogical approaches for sustainable development in higher education – a literature review and framework proposal. *Sustainability*, vol. 9, no. 11, 1889.

Redman, A., Wiek, A., & Barth, M. (2021). Current practice of assessing students' sustainability competencies – a review of tools. *Sustainability Science*, vol. 16, pp. 117-135.

Redman, A., & Wiek, A. (2021). Competencies for advancing transformations towards sustainability. *Frontiers in Education*, vol. 6, 785163.

Rieckmann, M. (2018). Learning to transform the world – key competencies in ESD. In: Leicht, A., Heiss, J. and Byun, W.J. (Eds.). Issues and Trends in Education for Sustainable Development. UNESCO, Paris, pp. 39-59. http://unesdoc.unesco.org/images/0026/002614/261445E.pdf

Salovaara, J. J., Soini, K., & Pietikäinen, J. (2020). Sustainability science in education – analysis of master's programmes' curricula. *Sustainability Science*, vol. 15, no. 3, pp. 901–915.

Sandri, O., Holdsworth, S., & Thomas, I. (2018). Assessing graduate sustainability capability post-degree completion. *International Journal of Sustainability in Higher Education*, vol. 19, no. 1, pp. 2–14.

Scherak, L., & Rieckmann, M. (2020). Developing ESD competences in higher education institutions – staff training at the University of Vechta. *Sustainability*, vol. 12, no. 24, 10336.

Trencher, G., Vincent, S., Bahr, K., Kudo, S., Markham, K., & Yamanaka, Y. (2018). Evaluating core competencies development in sustainability and environmental master's programs – an empirical analysis. *Journal of Cleaner Production*, vol. 181, pp. 829-841.

UNESCO (2017). Education for Sustainable Development Goals – Learning Objectives. Paris: UNESCO. http://unesdoc.unesco.org/images/0024/002474/247444e.pdf

United States National Academies of Sciences, Engineering, and Medicine (2020). Strengthening Sustainability Programs and Curricula at the Undergraduate and Graduate Levels. Washington, DC: The National Academies Press.

Weiss, M., Barth, M., Wiek, A., & von Wehrden, H. (2021). Drivers and barriers of implementing sustainability curricula in higher education – assumptions and evidence. *Higher Education Studies*, vol. 11, no. 2, pp. 42-64.

Wiek, A., Withycombe, L., & Redman, C.L. (2011a). Key competencies in sustainability – a reference framework for academic program development. *Sustainability Science*, vol. 6, no. 2, pp. 203-218.

Wiek, A., Withycombe, L., Redman, C.L., & Banas Mills, S. (2011b). Moving forward on competence in sustainability research and problem solving. *Environment – Science and Policy for Sustainable Development*, vol. 53, no. 2, pp. 3-12.

Wiek, A., Bernstein, M., Foley, R., Cohen, M., Forrest, N., Kuzdas, C., Kay, B., & Withycombe Keeler, L. (2016). Operationalising competencies in higher education for sustainable development. In: Barth, M., Michelsen, G., Rieckmann, M., Thomas, I. (Eds.) (2016). Handbook of Higher Education for Sustainable Development. Routledge, London. pp. 241-260.

Wiek, A., & Lang, D.J. (2016). Transformational sustainability research methodology. In: Heinrichs, H., Martens, P., Michelsen, G., & Wiek, A. (Eds.) (2016). Sustainability Science – An Introduction. Berlin, New York: Springer. pp. 31-41.

Wilhelm, S., Förster, R., & Zimmermann, A. (2019). Implementing competence orientation – towards constructively aligned education for sustainable development in university-level teaching-and-learning. *Sustainability*, vol. 11, no. 7, 1891.

A Rounder Sense of Purpose: Competences for Educators in Search of Transformation

5

Rick Millican

Abstract

After introducing the Rounder Sense of Purpose (RSP) project and its links to the United Nations Sustainable Development Goals (SDGs), this chapter introduces the 12 RSP competences, explaining the provenance and significance of each and why this combination might be considered transformational. The author goes on to explore some of the challenges related to adopting a competence-based approach, noting these challenges as: the concept of competence itself, presentation of the framework, pedagogy, outcomes and assessment. The concluding discussion on assessment is developed in a subsequent chapter in this book.

Keywords

Education for Sustainability (ESD) · Competences · Sustainable Development Goals (SDGs) · A Rounder Sense of Purpose (RSP)

Introduction

This chapter introduces the Rounder Sense of Purpose (RSP) framework while explaining the rationale behind the design. Links to the Sustainable Development Goals (SDGs) are explored, before reflecting on some key issues and challenges that have arisen for the project team as they have worked with these competences. This is followed by thoughts on future directions.

By drawing from and building on previous initiatives, principally the UNECE (2012) framework, *A Rounder Sense of Purpose* (RSP), an EU-funded project, has attempted to create a workable framework of competences for educators of sustainable development that could be employed in any sector or educational context. In combination, the competences cover those aspects of education highlighted by a broad body of research (e.g. Sleurs 2008; Wiek et al. 2011; Rieckmann 2012; Bertschy et al. 2013; Roorda 2016; Glasser and Hirsh 2016; Lozano et al. 2017), which have the potential to create learners who become active change agents working towards a sustainable future.

The project consisted of two phases: the first, RSP I, analysed the competences in the *Learning for the Future* framework (UNECE 2012), looking for overlap and redundancies and distilling the competences to a core of 12. While project partners tested and refined these competences in

R. Millican (✉)
University of Gloucestershire, Cheltenham, UK
e-mail: rmillican@glos.ac.uk

P. Vare et al. (eds.), *Competences in Education for Sustainable Development*, Sustainable Development Goals Series, https://doi.org/10.1007/978-3-030-91055-6_5

practice (working with a combined total of approximately 500 teachers, student teachers, lecturers and community educators), efforts continued to compare and contrast this emerging framework with other education for sustainable development (ESD) competence frameworks. This work assisted in defining three learning outcomes and a number of underpinning components for each of the 12 competences. This was then tested through a Delphi research process to check coverage (Vare et al. 2019). A second project phase refined the framework further and linked it with the Sustainable Development Goals (SDGs), producing materials to address both the SDGS and the competences, all freely available via the RSP website: https://aroundersenseofpurpose.eu.

The Rounder Sense of Purpose Framework

The name of the project—and resulting framework—was chosen to convey a broader purpose of education away from the tendency to focus on narrow economic imperatives, a trend that has been apparent for decades (Schultz 1960). A 'rounder sense' implies a more humanistic agenda of individual, collective and environmental well-being involving the emancipation and conscientisation of learners (Freire 2005). As a consequence, learners should be able to critique societal structures and instigate and participate in change in pursuit of a socially just and sustainable world.

The framework was displayed initially in a grid (see Table 5.1) with three columns, partly in recognition of its UNECE heritage but also as the three headings: thinking holistically, envisioning change and achieving transformation, encapsulated the essence of ESD capability (Frisk and Larson 2011).

Thinking Holistically

This cluster of four competences (the left-hand side column) reflects the fundamental understanding that all things are linked together in some way. The specific competences are:

Systems thinking: The need for learners to be able to understand systems and see the world as an interconnected whole, appreciating the connections between human and natural environments and recognising the consequences of actions taken and the causes of unsustainability (Jucker 2020).

Attentiveness: This refers to learners being attuned to what is happening in the world and attentive to, and aware of, aspects of human endeavour that are unsustainable and therefore requiring change. It is evident that technology and behaviours are constantly evolving and that research reveals some of these to have a negative impact on planetary systems (Bendell 2018). Consequently, learners need to have developed an interest in, and an ability to keep abreast of, new developments and emerging trends.

Transdisciplinarity: To assist with holistic thinking, learners need to appreciate the complexity of the issues involved, the limits of discipline-based knowledge and the dangers of adopting a single perspective (Selby and Kagawa 2010). To enrich and strengthen thinking, they need to be able to act and work collaboratively, engaging with others with different perspectives, including those outside of academic disciplines (in some contexts characterised as indigenous knowledge) in order to explore and construct new knowledge and ideas.

Criticality: This emphasises the need to constantly assess and evaluate ideas, knowledge and information with the willingness to challenge claims, opinions and assumptions (Sterling et al. 2017; Lotz-Sisitka et al. 2015). It recognises that knowledge is contingent on time, culture, power, evidence and perspective and that in order to progress towards a sustainable world it is necessary to separate fact from opinion and to question unsubstantiated claims.

Envisioning Change

These four competences—presented in the central column of the framework—acknowledge that

Table 5.1: The Rounder Sense of Purpose Framework

Thinking holistically	Envisioning change	Achieving transformation
Integration:		
Systems	Futures	Participation
The educator helps learners to develop an understanding of the world as an interconnected whole and to look for connections across our social and natural environments and consider the consequences of actions	The educator helps learners to explore alternative possibilities for the future and to use these to consider how behaviours might need to change	The educator helps learners to contribute to changes that will support sustainable development
Involvement:		
Attentiveness	Empathy	Values
The educator helps learners to understand fundamentally unsustainable aspects of our society and the way it is developing and increases their awareness of the urgent need for change	The educator helps learners to respond to their feelings and emotions and those of others as well as develop an emotional connection to the natural world	The educator develops an awareness among learners of how beliefs and values underpin actions and how values need to be negotiated and reconciled
Practice:		
Transdisciplinarity	Creativity	Action
The educator helps learners to act collaboratively both within and outside of their own discipline, role, perspectives and values	The educator encourages creative thinking and flexibility within their learners	The educator helps learners to take action in a proactive and considered manner
Reflexivity:		
Criticality	Responsibility	Decisiveness
The educator helps learners to evaluate critically the relevance and reliability of assertions, sources, models and theories	The educator helps learners to reflect on their own actions, act transparently and to accept personal responsibility for their work	The educator helps learners to act in a cautious and timely manner even in situations of uncertainty

the society we have created is currently unsustainable and that we need to imagine alternative ways of being that can contribute to a sustainable future. The group comprises:

Futures: This refers to being able to project into the future and, by looking at current trends, predict what may happen. This trend or pattern analysis is necessary in order to anticipate and therefore mitigate, or at least prepare for, challenges that are likely to occur. It also refers to the ability to imagine alternative futures which are more sustainable and to consider what steps are needed to be taken to reach these preferable visions of the future (Rieckmann 2012).

Empathy: At one level this involves seeing the world from another's perspective, whether that is as another person or a different being altogether (Kopnina and Cherniak 2016). This competence also recognises that exploring future scenarios can challenge both ourselves and others. It requires awareness of one's emotional response to the threats that we face and an understanding of how others may react. This may involve protecting each other and enhancing resilience given that a loss of hope can be damaging to mental health and impact our motivation and ability to work towards achieving change (Ojala 2016).

Creativity: The process of envisioning alternative futures demands creativity—the ability to imagine different ways of being and behaving, of ways we might structure and organise society, of ways we could structure the economy and of ways to design and build items we need including the buildings we live and work in. It also requires flexibility, the willingness to accept change where appropriate, to revisit tried and tested ideas, possibly in new combinations, as well as a readiness to try the new (Carrascal et al. 2019).

Responsibility: The final competence in this column focuses on the need to accept responsibility for things we do and the decisions we make. Linked to futures thinking, this competence focuses on who we are—the *Learning to Be* dimension (UNECE 2012). It encourages consideration of the consequences of actions taken, including the educator's inevitable role in providing a model that their students may follow.

Achieving Transformation

This final cluster suggests that, having thought holistically, imagined the future and recognised the need to change, it is necessary to think about the moves required to make change happen. Competences in this group are:

Participation: This involves understanding that there are different ways to participate as well as varying levels of participation and that each may be valid in different contexts. Working together, deliberating on decisions with others; these are key elements of a democratic response to our current unsustainable predicament (Reid et al. 2007).

Values: This recognises that actions are underpinned by beliefs about how the world should be, which are themselves based on values. The competence refers to the ability to recognise and understand values as drivers behind the behaviours of self and others, what Barth et al. (2007) call value interiorisation. It encompasses the need to acknowledge and accept differing value systems and the need to be willing to share, discuss and reflect upon these differences in the context of sustainability.

Action: Ultimately, the process of achieving transformation requires action and this competence refers to the ability to act. It is about the development of agency and having the confidence and skills required to be able to take action in different contexts making use of social, political, economic and democratic structures (Ploum et al. in Carrascal et al. 2019).

Decisiveness: This refers to the ability to make decisions, particularly when faced with dilemmatic situations that are so characteristic of sustainable development. It recognises that knowledge and understandings about the world in general and sustainability in particular are incomplete and evolving and that therefore there is a need to have the confidence to make considered decisions based on the information to hand (Vare 2019).

Links to the UN Sustainable Development Goals (SDGs)

As a Contribution to SDG 4: Quality Education

Sustainable Development Goal 4 states that education 'enables upward socioeconomic mobility and is a key to escaping poverty'. Whilst there is likely to be little dispute about the need to widen access to education to include all and that it can help escape poverty, there is some contention about the broader purpose, content and method of education (Curtis and Pettigrew 2009) and what constitutes 'quality' education (Kumar and Sarangapani 2004). Young people increasingly want an education that will inform them about the challenges facing the world and equip them with the skills that will help them tackle these issues and mitigate them where possible (e.g. https://www.teachthefuture.uk/). SDG 4 therefore calls explicitly for education for sustainable development (Target 4.7).

As Schumacher (1997) said while observing how an increase in the volume of education had been accompanied by an increase in the danger of ecological collapse, 'if still more education is to save us, it would have to be education of a different kind' (ibid). An educator equipped with the RSP competences should be well placed to deliver a 'different kind' of education, one that should go further in satisfying young people's demands to be prepared for the future.

As a Means to Address Other SDGs

The 17 Sustainable Development Goals are billed as a 'shared blueprint for peace and prosperity for people and the planet, now and into the future' (https://sdgs.un.org/goals) and note the parallel aims of ending poverty, improving health and tackling climate change (ibid). Conceptions of sustainable development recognise the need to consider societal and economic needs alongside environmental needs and stress the impossibility of achieving environmental sustainability without economic models and societal structures which are themselves sustainable and designed in harmony with the environment (Purvis et al. 2019). It can therefore be seen that utilising the RSP competences and educating towards sustainable development is in keeping with educating towards the SDGs.

Consequently, a mapping exercise was undertaken showing this interplay between the RSP framework and the SDGs. Activities and materials were produced for educational contexts illustrating how both could be covered simultaneously (see https://aroundersenseofpurpose.eu/sdgs/).

Key Issues/Challenges

In the process of designing and articulating the RSP framework in terms of competences, a number of issues have arisen which the project team has wrestled with. Five of the key issues are discussed below.

The Concept of Competence

The move to an expression of educational aims in competence terms rather than descriptions of content was an attempt to shift focus from what should be conveyed or transmitted to the learner, to expectations as to what the student should be able to do, "the output-oriented approach focuses on students gaining the 'concepts and abilities for social action'" (Frisk and Larson 2011: 6). However, what initially may have appeared to be a simple task, proved complicated by considerations of level and context. The question remains as to whether it is possible to be competent in one context but not in another and whether there is a baseline or threshold competence and thereafter different levels of capability that would need to be defined (e.g. Roorda 2019).

There were also different interpretations of competence across Europe, ranging from the more prescriptive, rigid, skills-based perspective of the UK, to a looser, more developmental perspective found in some other partner countries (Shephard et al. 2018; see also the Introduction to this book).

In addition, it was clear that demonstration of a competence involved a set of other subskills, knowledge and attitudes and a realisation that a full articulation involves a level of complexity that can be confusing and off-putting for users and run counter to the systemic and holistic way of thinking that characterises so much of sustainability discourse (Vare et al. 2019).

For educators, the dilemma remains as to where to focus their efforts—the act of doing, the underpinning and related knowledge, or the values conveyed. One might wonder whether having a commitment to sustainability with its implicit values is a pre-requisite to achievement, or whether being able to act in an appropriate way is sufficient. Shephard (see Chap. 6) argues that, in addition to having the capability and the possession of sustainability values, there should also be a willingness or motivation to act. This in turn raises issues in terms of assessment, particularly in defining acceptable means for students to demonstrate their intent (see Chap. 21).

In confronting these issues, the project team adopted a pragmatic approach. The competences were kept to a core 12 in order to produce something useable and achievable, but with lists of learning outcomes and underpinning knowledge for those who were looking for more depth and detail.

The level of competence was also left open to interpretation in recognition of the varying contexts in which they were to be applied, with some partners aiming them at International Standard Classification of Education (ISCED) Level 3 and others at Level 7. Although expressed as competences, they were also viewed as developmental and progressive.

In addition, it was felt that although certain values were desirable, it was perhaps unethical and inappropriate to demand an educator to necessarily hold a prescribed set of values. An analogy could be that of teaching religious education, whereby it is possible to teach about different belief systems, without necessarily holding (or promoting) a faith of one's own. This in turn challenges the *Responsibility* competence, which calls on educators to be role models.

Graphic Representation

Having established a core set of 12 competences that aligned with the columns of the UNECE framework, RSP was portrayed initially using a grid layout as shown above. This was convenient as it hinted at its heritage, had a presentational simplicity and was helpful for course design and potentially assessment plans.

The danger with adopting such a design however, is that it can lead to a linear interpretation with trainers and students tending to adopt a reductionist approach to the competences, considering them as individual and discrete items to be viewed and used one at a time.

This runs counter to the philosophy behind the idea of the educator as a system with the various competences acting and interacting together in different combinations according to context. In an attempt to overcome this, the RSP competences are depicted on an artist's palette (Fig. 5.1) with the educator mixing and using them as required to suit their context.

Pedagogical Approach

The vocational background to competences has the potential to suggest a 'skills-training' approach to implementation as one might, for example, teach machinery operation or booking-in procedures at a hotel. This conveys the image of a technocratic, task analysis exercise whereby the whole process is divided into steps to be followed in which the student is then trained. This is not in keeping with most perceptions of the teaching process, and of education for sustainability in particular, which tend to favour a social-constructivist, critical pedagogy in which knowledge is co-constructed and which is designed to empower and develop agency and independence of thought.

This issue remains unresolved within the project. Activities supplied as examples of how the competences could be developed largely imply a constructivist pedagogical approach, but there is still the question as to whether that approach should be articulated explicitly as 'the way to do this'.

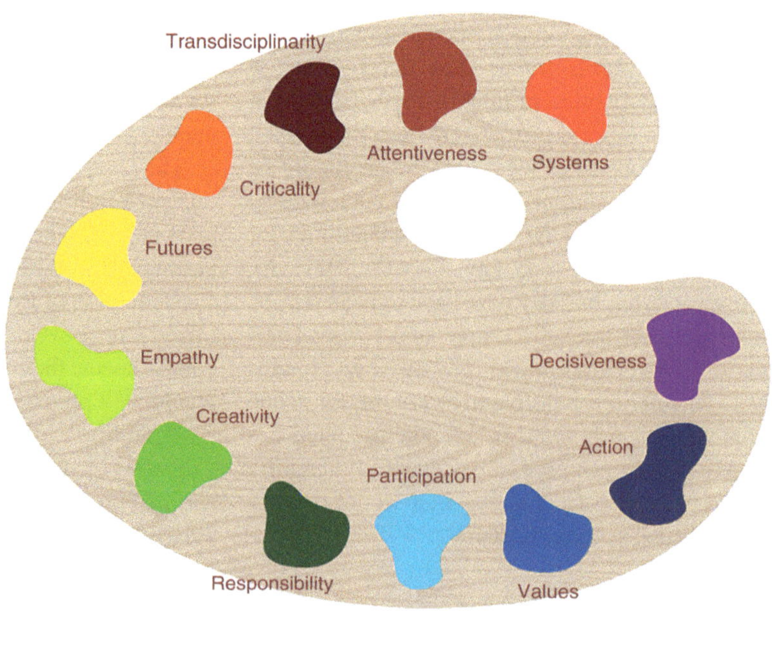

Fig. 5.1 The RSP Palette

Learning Outcomes

Despite—or perhaps because of—the competence framework having a strong heritage and having been tested rigorously, it offers a set of outcomes prescribed by 'experts'. The power and control over what is to be learned is therefore determined by someone other than the learner.

This is not in the spirit of the critical pedagogy that the project team preferred, in fact it is more in line with a traditional, transmissive 'banking' approach (Freire 2005) to education. Ultimately, this was deemed acceptable because the competences, although predetermined, were expressed simply as a range of capabilities rather than a prescription of how and when these should be applied.

Assessment

The challenge of assessing competences has been reported on elsewhere (see, for example, Redman et al. 2020) with now familiar questions as to what evidence is appropriate (e.g. witness statement, photos, journals, reflections), how much is enough (evidence of all components needed?) and how level is determined (see Chap. 21).

Partners have responded to this challenge in different ways depending on context, need and level hinting at the absence of a perfect, transferrable method. Example approaches can be found on the website https://aroundersenseofpurpose.eu/.

Another issue arises when considering what is actually being assessed. Given that ESD is concerned with transformation, is demonstration of the competences sufficient or should we be looking for evidence of transformation and if so, of the educator, the educator's learners or of the learners' impact on society? This in turn relates back to the issue mentioned earlier as to whether assessment should be of the learner's ability to perform, the intent or values behind their abilities and/or their motivation and willingness to act (Chap. 6).

Where Next?

At the time of writing the project is nearing the end of its second phase and approaching successful completion of its objectives, i.e. an ESD competence framework with links to the SDGs, a full set of example activities that address the competences and SDGs, a website and publication of a range of academic papers and a (this) book about competences.

The hope is that the framework will be used increasingly to help develop educators who will, in turn, educate people for *a rounder sense of purpose*, in-tune with the needs of the planet and of the need to create change to help meet those needs.

There is much work still to be done in equipping people with the knowledge, skills and values necessary to participate in working towards a sustainable world. It is hoped that the RSP framework can play a significant part in clarifying what it takes to be an educator who can help develop such people.

Project Partners

Duurzame PABO, The Netherlands (RSP I and II).

Frederick University, Cyprus (RSP I and II).

Haute Ecole Pedagogique du canton de Vaud, Switzerland (RSP II).

Italian Association for Sustainability Science, Italy (RSP I and II).

Kutato Tanarok Orszagos Szovetsege, Hungary (RSP I and II).

Tallinna Ulikool, Tallinn, Estonia (RSP I).

University of Gloucestershire, UK (project lead RSP I and II).

Universititat Oberta de Catalunya, Spain (RSP II).

University of Vechta, Germany (RSP II).

References

Barth, M., Godemann, J., Rieckmann, M. and Stoltenberg, U. (2007). 'Developing key competencies for sustainable development in higher education'. International Journal of Sustainability in Higher Education, 8 (4), pp. 416–430

Bendell, J. (2018) Deep Adaptation: A Map for Navigating Climate Tragedy. IFLAS Occasional Paper 2

Bertschy, F., Künzli, C. & Lehmann, M. (2013) Teachers' Competencies for the Implementation of Educational Offers in the Field of Education for Sustainable Development. *Sustainability*, 5, 5067–5080, doi:https://doi.org/10.3390/su5125067.

Carrascal, G, Magro, M., Anguita, J. & Espada, M. (2019). Acquisition of Competences for Sustainable Development through Visual Thinking. A Study in Rural Schools in Mixco, Guatemala. *Sustainability, 11*, 2317

Curtis, W. & Pettigrew, A. (2009). *Learning in Contemporary Culture.* Exeter: Learning Matters

Freire P (2005) *Pedagogy of the Oppressed.* New York: Continuum

Frisk, E., & Larson, K. (2011). Educating for Sustainability: Competences & Practices for Transformative Action. *Journal of Sustainability Education.* Vol 2. ISSN: 2151-7452

Glasser, H. & Hirsh, J. (2016). Toward the Development of Robust Learning for Sustainability Core Competencies. *Sustainability, 9(3).* 121-134 DOI: https://doi.org/10.1089/sus.2016.29054.hg

Jucker, R. (2020) *Can We Cope with the Complexity of Reality: Why Craving Easy Answers Is at the Root of our Problems.* Newcastle, UK: Cambridge Scholars Publishing

Kopnina, H. & Cherniak, B. (2016). Neoliberalism and justice in education for sustainable development: a call for inclusive pluralism, Environmental Education Research, 22:6, 827-841, DOI: https://doi.org/10.1080/13504622.2016.1149550

Kumar, K. & Sarangapani P (2004). "Paper commissioned for the EFA Global Monitoring Report 2005, The Quality Imperative.

Lotz-Sisitka, H., Wals, A. E., Kronlid, D., & McGarry, D. (2015). Transformative, transgressive social learning: rethinking higher education pedagogy in times of systemic global dysfunction. *Current Opinion in Environmental Sustainability* 16, 73-80.

Lozano, R., Merrill, M.Y., Sammalisto, K., Ceulemans, K. & Lozano, F. J. (2017). Connecting Competences and Pedagogical Approaches for Sustainable Development in Higher Education: A Literature Review and Framework Proposal. *Sustainability* 2017, *9*, 1889-1904; doi:https://doi.org/10.3390/su9101889

Ojala M (2016). Hope and anticipation in education for a sustainable future in *Futures* 10:04 https://doi.org/10.1016/j.futures.2016.10.004

Purvis, B., Mao, Y. & Robinson, D. (2019). Three pillars of sustainability: in search of conceptual origins. *Sustain Sci* **14,** 681–695 https://doi.org/10.1007/s11625-018-0627-5

Redman, A., Wiek, A. & Barth, M (2020). Current practice of assessing students' sustainability competencies: a review of tools Sustainability Science https://doi.org/10.1007/s11625-020-00855-1

Reid, A.D., Jensen, B.B., Nikel, J. & Simovska, V. (Eds.) (2007). *Participation and Learning: perspectives on education and the environment, health and sustainability.* Dordrecht: Springer Press

Rieckmann, M., (2012). Future-oriented higher education: Which key competencies should be fostered through university teaching and learning? *Futures.* 44 127- 135

Roorda, N. (2016). The seven competences of a sustainable professional. The RESFIA+D model for HRM, education and training. Chapter 1 in *Management for Sustainable Developmen*t, Carolina Machado & J. Paulo Davim (eds.), River Publishers, Aalborg, Denmark, 2016, www.riverpublishers.com/book_details.php?book_id=348

Roorda, N. (2019). The Seven Sustainability Competences according to the RESFIA+D Model. *The Central European Review of Economics and Management* 3(3):45-87 DOI: https://doi.org/10.29015/cerem.781

Schumacher, E. F. (1997). *'This I believe' and other essays* (essay first published in 1974). Dartington: Green Books.

Selby, D., & Kagawa, F. (2010). Runaway Climate Change as Challenge to the 'Closing Circle' of Education for Sustainable Development. *Journal of Education for Sustainable Development*, *4*(1), 37–50. https://doi.org/10.1177/097340820900400111

Shephard, K., Rieckmann, M. & Barth, M. (2018). Seeking sustainability competence and capability in the ESD and HESD literature: an international philosophical hermeneutic analysis. In: Environmental Education Research. First Online: 24 October 2018. DOI: https://doi.org/10.1080/13504622.2018.1490947.

Schultz, T.P. (1960). Capital Formation by Education. *Journal of Political Economy* 68(1) pp571-583

Sleurs, W. (2008). *Competencies for ESD (Education for Sustainable Development) Teachers: A Framework to Integrate ESD in the Curriculum of Teacher Training Institutes; Curriculum, Sustainable Development, Competences, Teacher Training (CSCT);* Comenius 2.1 Project; UN: Brussels, Belgium, 2008; Available online: https://www.unece.org/fileadmin/DAM/env/esd/inf.meeting.docs/EGonInd/8mtg/CSCT%20Handbook_Extract.pdf (accessed 27 November 2020).

Sterling, S., Glasser, H., Rieckmann, M., & Warwick, P. (2017). 'More than scaling up': a critical and practical enquiry into operationalising sustainability competences. Chapter 10 in Corcoran, P., Weakland, J., Wals, A. (eds) *Envisioning futures for environmental and sustainability education.* Wageningen: Wageningen Academic Publishers

UNECE (2012). *Learning for the Future. Competences in Education for Sustainable Development*; UNECE: Geneva, Switzerland, 2011; Available online: https://www.unece.org/fileadmin/DAM/env/esd/ESD_

Publications/Competences_Publication.pdf (accessed on 21/12/2020).

Vare, P. (2019). Beyond the 'green bling': Identifying contradictions encountered in school sustainability programmes and teachers' responses to them. *Environmental Education Research*, 26:1, 61-80, DOI:https://doi.org/10.1080/13504622.2019.1677859

Vare, P.; Arro, G.; de Hamer, A.; Del Gobbo, G.; de Vries, G.; Farioli, F.; Kadji-Beltran, C.; Kangur, M.; Mayer, M.; Millican, R.; Nijdam, C.; Réti, M.; Zachariou, A. (2019). Devising a Competence-Based Training Program for Educators of Sustainable Development: Lessons Learned. *Sustainability*, *11*, 1890. https://www.mdpi.com/2071-1050/11/7/1890

Wiek, A., Withycombe. L., & Redman, C. (2011). Key competences in sustainability: a reference framework for academic programme development. Sustain Science, 6:203–218 DOI https://doi.org/10.1007/s11625-011-0132-6

6 On the Educational Difference Between Being Able and Being Willing

Kerry Shephard

Abstract

This chapter questions the role of competence in the ESD mission. It explores how competence is defined and used to describe intended outcomes in ESD and argues that definitions of competence widely used in the ESD discourse conflate cognitive and affective objectives, relatable to being able to do something and being willing to do the same thing, in a way that is unhelpful in the contexts of teaching and learning. The chapter suggests that teaching a student to be able to perform a sustainability-related behaviour and teaching that same student to be willing to perform that behaviour are different educational tasks, requiring different teaching and learning approaches, different assessments, and perhaps even different teachers. The chapter submits that replacing the term *competent* with the terms *able* and *willing*, as appropriate, would allow ESD practitioners to communicate with one another and others far more effectively than they do at present.

Keywords

Competence · Able and willing · Dispositions · Higher education for sustainable development (HESD)

Introduction

Education for sustainable development (ESD), in general, and for higher education in particular (HESD), has many expectations. These involve knowledge and skills (relating to being able, or capable, to behave in ways compatible with sustainability and the sustainable development goals, and so relating to learning objectives that we can readily describe, teach, and assess) but also more challenging expectations, for example, that our learners will be willing to behave in the world in a different way from how previous generations have behaved that are values-based and that may not sit easily within some educational discourses. Terms like competence[1] and disposition have found their way into the vocabulary of ESD and sustainability science with broad meanings that conflate being able and willing and that apparently make pedagogical sense to practitio-

K. Shephard (✉)
Higher Education Development Centre, University of Otago, Dunedin, New Zealand
e-mail: kerry.shephard@otago.ac.nz

[1] Competence is used as a mass noun in this chapter. Competences, competency, and competencies are also used where necessary to comment on the work of others.

P. Vare et al. (eds.), *Competences in Education for Sustainable Development*, Sustainable Development Goals Series, https://doi.org/10.1007/978-3-030-91055-6_6

ners in these disciplines, including I assume, some other contributors to this book (although not, I hasten to add, to me). This chapter is also, to a degree, about the competence and dispositions of critical thinking (as something that most educators agree contributes to the link between learning and behaviour). Critical thinking does involve using a range of cognitive skills, or abilities, but is also dependent on a range of affective dispositions that underpin these abilities. These dispositions, or affective attributes, for example, help a person to decide whether or not to bother thinking critically, for some personal effort is always involved in critical thinking.

What Does It Mean to Be Competent?

Our exploration needs to start with a frequently communicated and likely widely accepted current definition of competency, in the context of ESD, as provided by Brundiers et al. 2020, and previously by Rieckmann (2012); '*Competency; Cluster of specific and interrelated individual dispositions comprising knowledge, skills, motives, and attitudes,* i.e., *combining cognitive, affective, volitional and motivational elements. Competency facilitates self-organized action, a pre-condition to achieve successful performance and a positive outcome in various complex situations, responding to the specific situation and context*' (np). Abilities and willingness to perform these abilities are both explicit within this definition, with its clear commentary on successful performance and positive outcomes. With respect to the decisions that competent people make there are commonalities between this use of the term competency and that developed by Wiek et al. (2011), who included normative competency within their list of key competencies, suggesting that; '*The concept of sustainability is unavoidably value laden and normative, since it addresses the question of how social-ecological systems ought to be developed, so that they balance and even enhance socio-economic activities and environmental capacities...*' (p. 209). Strictly speaking the Brundiers et al. 2020 definition leaves the hard work of definition to another word, disposition. Competency in this respect is merely the collective noun for the many dispositions involved. Dispositions are imbued with the wide diversity of personal attributes and educational outcomes, both cognitive and affective, required by any individual who is to be both able and willing: knowledge, skills, motives, and attitudes. In this context, dispositions may be essentially latent, perhaps even unobservable, but become apparent, and even measurable, when integrated into a competence. Such meanings can be traced backwards and forwards throughout the ESD literature. A recent exploration of links between teaching approaches and outcomes (intended and achieved) in the context of teacher education by Brandt et al. (2021) identifies; '*... the three competence aspects of content knowledge (CK), pedagogical content knowledge (PCK), and attitude...*' (p.2) combining both cognition and affect into the one entity of competency. Going backwards almost three decades, we should note that (in the context of sustainability-related education); "'*Competence' is associated with being able, and willing, to be a qualified participant*" (Jensen and Schnack 1997, p.165).

Overall, it would appear from this analysis that competent people in the context of ESD have knowledge, skills, motives, and attitudes to not only be able to perform if they happen to wish to do so, but be willing to do so as well, resulting in successful outcomes. Incompetent people (or more kindly, less-competent people) would lack some aspect of this competence. Perhaps they would have the required knowledge and cognitive skills to do the task if they chose to do it, but be unwilling to do so, perhaps lacking the right attitude, or motivation, or volition. In these terms, ESD clearly needs to ensure that they sufficiently learn the right attitude, motivation, volition, or emotional attachment, to the task at hand, to be willing to undertake it. No doubt that is why, for some, this is education *for* sustainability, not education *about* sustainability, or education *that might lead to* sustainability *if those being educated wanted it to.*

What Does Competent Mean in an Everyday Sort of Way and How Did Its Technical Meaning Change So Much?

In common English usage, and noting that competence has multiple and overlapping meanings, competence 'as ability' is rarely confused with the 'performance of that ability'. A quick search with Google for 'able and willing' should be enough to suggest to most readers that in common usage we need both words because they are two different things. When people do not have a choice we can get away with just one, as in the navy's Able Seaman, whose willingness is always assumed; but for citizens with free will, we need two. The Oxford English Dictionary (OED) does support this assertion, but also identifies that in some contexts, debate on ability and performance has occurred and that willingness to perform can creep into some meanings of competence. Conversations with colleagues from many countries over many years confirm that in other languages and cultures, common usage of words deemed to be equivalent to competence is often similarly straightforward, meaning ability, but that competence may also have several different meanings in most languages. On this basis, competence is not a good word to use in technical discourses, particularly those in international, multilingual contexts. A recent review and summary of competence, from several decades of international discourse in philosophy, psychology, linguistics and education, categorises competence variously as ability/capability (common in English), as disposition (common in German), as a process, as a relation, and as a quality or state of being (Schneider 2019). This particular theoretically grounded analysis does come to a particular conclusion about a fundamental meaning (competence as a cognitive ability, focusing in an educational-science context on the ability of a person to manage demands) but also notes limitations and domain specificity of the resulting definition. A recent similar analysis exploring the development and application of competence in educational theory in German and wider contexts, in the context of foreign language learning, describes why competence has come to dominate that discourse in recent years and emphasises limitations imposed by its broad and ill-defined nature and the importance of unpacking its contributory meanings (Glaesser 2019). Glaesser (2019), using a restricted ability-based conception of competence, suggests that; '*In practice, faced with a situated task, all of motivation, knowledge, and competence will be jointly required for the individual to be able to act successfully. But in order to unpack their respective contributions and importance, clear conceptual distinctions are needed*' (71). Glaesser (2019) referencing others, suggests, however, that clear conceptual distinctions have not necessarily been dominant in formulating educational policy in Germany, which currently emphasises educational outputs as competencies; '*In Germany, educational policy traditionally tended to be input-oriented, with little attention being paid to outputs. However, following the so-called PISA-shock,* i.e., *the results of the first Programme of International Student Assessment (PISA) administered by the OECD which showed Germany to be in a lower position in the international rankings than had been hoped and expected, educational policy and rhetoric in Germany changed (Ertl 2006)*' (72).

It should be interesting to note that links between competence and education go back a long way. Hodge (2007) elegantly argues that competency-based education (and similar multiple ways of integrating competency with learning and teaching) developed primarily in the USA was a response to political and social pressures, rather than theoretical necessity, and substantially as an element of a culturally based movement towards accountability, in particular of teachers and teacher-trainers, in the late 1950s. At that time, the USA was behind the USSR in the space race and politicians argued that the education system had to improve and the best minds of the time were put to the task. Competency-based education and training (CBET) developed in the USA as an extension of behaviourism, the dominant learning theory at the time. And in these contexts, competency explicitly included knowledge, skills, and behaviour, and to a significant degree assessment of competency involved assessment of behaviours. Clear evolutionary lines can be drawn between the development of CBET, of learning outcomes, of criterion-referenced assessment and of theoretical differentiation between education and training. Nowadays, of course, and especially in the context of ESD, we struggle to separate the education of professionals from that of other learners, we take what works from the training discourse and apply it to education, and we forget where these terms and concepts came from.

Perhaps one conclusion can be drawn from these analyses. In tough times some educators have felt the need to limit free choice and maximise the development of ability, in particular by minimising opportunities for learners to be unwilling; education becomes training, cognition subsumes affect and competence assumes multifaceted meanings that no one should argue about.

What Then of Dispositions?

In common English usage, dispositions are essentially emotional or affective personal characteristics. The OED lists several meanings, the nearest to our interest is 'Natural tendency or bent of the mind, *esp.* in relation to moral or social qualities; mental constitution or temperament; turn of mind' (OED 2021); clearly and essentially affective in nature and not combinations of cognition and affect. Whereas in the Brundiers et al. 2020 definition of competence, dispositions combine cognition (as knowledge and skill) and affect (as motive and attitude), in common English usage dispositions say more about what people might be willing to do with their abilities than they do about the abilities themselves. Internationally, and in our context, dispositions are every bit as troublesome as competence in attempting to combine what people are able to do, with what they are willing to do.

As with all affective attributes, it would not be reasonable to say that dispositions cannot or should not be taught. The OED definition of disposition does, however, indicate why the teaching of some dispositions (and their related affective values and attitudes) may be problematic for many teachers, who may be troubled to comprehend their role as changing the natural tendency of their pupils or students in relation to moral or social qualities, perhaps particularly if they do not share the moral or social qualities being taught or if pupils' existing qualities are broadly representative of societal norms. Shephard (2008), drawing on the work of Bloom et al. (1971), suggested that education has historically avoided such affective goals, being concerned about charges of indoctrination or brainwashing. Shephard (2008) encouraged ESD practitioners to use the Krathwohl et al. (1964) affective hierarchy to determine for themselves how far up the hierarchy they were prepared to teach and assess, and specifically to distinguish ability from willingness; '*... some teachers are apparently prepared to go further. They require students to develop particular attitudes and to behave in particular ways, often in relation to the stated values of their future profession, and assess them on their ability, and willingness, to do so*' (Shephard 2008, p.96). Those who integrate affective outcomes within learning objectives described as competence within the ESD discourse apparently are prepared to go all the way. Shephard (2020) provides a rationale for distinguishing some forms of affective outcomes that should be taught in higher education, from others that perhaps should not, and so an argument for why teachers at all levels need to think critically about this important distinction.

The relationship between competence, as elements of Bloom et al.'s cognitive hierarchy of knowledge and skills (Bloom et al. 1956) and dispositions, as expressions of Bloom et al.*'s* affective hierarchy of values, attitudes and behaviours (Krathwohl et al. 1964) has been implicit in the related discourse on critical thinking, integral to a range of disciplines including education. Advocates for critical thinking in education and in educational research would identify expert-group deliberations in the late 1980s on critical thinking skills and dispositions (see in particular, Facione 1990, 2000) as central to a wide range of disciplinary discourses. These discourses are generally characterised as accepting that cognitive and affective learning outcomes, while interacting, are different. They require different teaching and learning activities and means of assessment. Shephard (2008) and Shephard and Egan (2018) summarised the literature that has emphasised the critical importance of role models in educational endeavours that focus on learning in the affective domain.

Is Competence an Intolerably Blunted Conceptual Tool, in the Hands of ESD Practitioners?

Shephard and Brown (2016) asked how democratic HESD might be, as a means to explore how the word democracy was used and understood in this discourse. They concluded that democracy, in this context, may have become another of Sartori's '*intolerably blunted conceptual tools*' (Sartori 1970, 1053). Shephard et al. (2019) continued to ask if miscommunication or misunderstanding of concepts within ESD might be contributing to our slow progress. They concluded that; '*educational meanings of the widely used terms 'competence' and 'capability' are not adequately understood and shared to be helping ESD and HESD to progress*' (13). This chapter integrates greater understanding of the diverse meanings of competence afforded by Schneider (2019), Glaesser (2019) and Shephard (2020) to suggest that perhaps competence is problematic for ESD. But how problematic is competence, and in which circumstances?

Does Conflating Cognition and Affect Matter? Is This Mere Semantics?

It could be argued that all disciplines legitimately appropriate words from one or more languages and, through the complex processes of disciplinary discourse, create new meanings for these words that afford discipline-specific debate capable of moving the discipline as a whole onwards. It could even be that such appropriation is required in circumstances where new words or phrases cannot be imagined. Relevant here is the utility of using a word already in common usage and already capable of activating communication frames valuable to those who need to (Lakoff 2010). As suggested by Norris (1991, 331) 'The language of competency-based approaches to education and training is compelling in its common-sense and rhetorical force'.

Arguably it really does not matter if being willing and being able are combined into a single construct [let's call it an aptitude, as the *quality of being fit for a purpose or position*, although vocation, as in a *strong feeling of suitability for a particular career or occupation*, may also be relevant here] if those who possesses this quality, feeling, property, or construct have limited personal choice whether to make use of it or not. Professionals, in general, willingly give up some personal freedoms in order to be members of a profession. Some do not need to give anything up, having acquired the quality, feeling, property, or construct in abundance through earlier education, critical incident, or social upbringing. Some early work on competence relating to sustainability was clearly in the context of sustainability professionals. Wiek et al. (2011) were so targeted, and this approach continues in the work of Brundiers et al. (2020) and perhaps Evans (2019). Whether these professionals are willing to put their abilities to use is not necessarily an important question, if having joined the profession they do not personally need to make choices in these regards. Conflating being willing and being able is hardly objectionable in this context, albeit unfortunate.

A similar situation may exist in certain circumstances in relation to other roles that historically have not necessarily needed to embrace sustainability, but in future may be expected to. My long personal experience with the NEP (Revised New Ecological Paradigm scale) assures me that within every group of university biology students, for example, a few will score straight 5 s. In my hands this tool is always used with anonymous respondents, so I do not know which in the group are Eco-warriors (although generally to be fair and to their credit they do make themselves known), but I am confident that for these individuals, conflating knowledge, skills, and affect poses no problem. Assuming that their values are matched by their sustainability knowledge and skills, these students will have at least some ESD competence in abundance.

And, to be blunt, conflating affect and cognition into the catch-all, frame-activating, educationally trendy world of competence really would not matter much to me, or to anyone else probably, if it worked. If in 2021 our ESD field of enquiry was congratulating itself on a job well done, this chapter likely would not find a place in this book.

When It Really Does Matter

The problem occurs when these same properties are applied conceptually or practically to those who have not chosen to adopt the values of any particular profession, and for whom the label of incompetent is hardly appropriate if all they have done is chosen not to be willing to put their learned knowledge and skills to use. As Glaesser (2019) states, '*If a concept is too broad—for example, if competence is taken to include motivational states—it is harder to study and may be less suited to explaining the outcome of interest*' (p.71). It really does matter to the broader field of enquiry identified as ESD that we separate being able from being willing, especially in educational contexts such as those described for educational programmes in general by UNESCO (2017). Perhaps, after all, there really is a difference between training and education.

I was thinking about this when I read the account of recent research (Brandt et al. 2021) into learning in a compulsory sustainability course for trainee teachers in the USA. I noticed in particular that these trainee teachers on average started off with a mean NEP score of 3.71 and ended with a marginally higher mean NEP score of 3.92. I had several emotional responses to this data. I wondered if even though their university teachers had conceptualised their sustainability learning as competence and conflated cognition and affect within their notion of competence, the students might not have done so. Noting that means hide all sorts of interesting data, on average these students may not have been more inclined towards facets of sustainability (in an affect, values, dispositional sense) at the end of the course, than they were at the beginning. I was also pleased to read in the conclusion section of this paper how important it is for teachers in training to experience role models for sustainability, rather than, presumably, university teachers who teach sustainability but do not necessarily role model it. But overall, I did wonder if these teachers, fully trained (or educated?), might be able to teach their pupils to be able to address sustainability demands, but less able to teach their pupils to be willing to do so. Indeed, I doubt that these teachers, on average, will be willing to attempt to do so themselves, unless of course they become contractually obliged to try. (Perhaps hidden within the mean NEP score of 3.92 are some high scoring eco-warriors; I hope so). I have similar emotional responses to the Rounder Sense of Purpose project (https://aroundersenseofpurpose.eu), described in Chap. 5 of this book, given its role in supporting schoolteachers to teach sustainability. Do we really want able teachers (in the sense of their knowledge and skills) teaching our young people anything to do with sustainability *unless* they choose to role model what it is that they are obliged to teach? If they do not, might they do more harm than good, with respect to the mission and obligations of compulsory education in most nations nowadays? Might we be contributing to a generation of knowledgeable bystanders to social, economic, and environmental collapse? Perhaps the struggles of Redman et al. (2020) to link ESD competencies to assessments, and the diversity of linkages between sustainability and transformative learning identified by Rodríguez Aboytes and Barth (2020), and the limited success of Scherak and Rieckmann's (2020, p.1) efforts to use '*staff training*' to '*enable in-service and pre-service educators to demonstrate their competence in ESD*' all illustrate the dangers of using a blunted conceptual tool in the development of our field of enquiry. I made the mistake once of commenting that teachers committed to teaching sustainability needed to cycle to school and wear shabby recycled clothes, to be effective role models for sustainability. A colleague corrected me, noting that recycled clothes are not necessarily shabby.

Is There Some Good News Here?

In my many discussions on these issues with many colleagues from many different nations and cultures, in amongst total disarray about the use of competence, I have yet to find disagreements about the meanings of the words 'able' and 'willing' or their equivalents in any language. Such broad agreement provides hope that our ESD field of enquiry can make clearer statements about what teachers teach and assess, with clear statements about educational objectives pertaining to particular educational levels and jurisdictions, and reasonable ways to match teachers to objectives. By choosing ESD competence as an educational objective and measuring stick, academics in the broad areas of environmental education, education for sustainability, and education for sustainable development appear to me to have made a simple situation far too complicated.

Perhaps those who teach sustainability in compulsory education, supported by the educational strategies of their local governments that insist that they teach their pupils to *be* sustainable will be mindful to consider what it is they are attempting and how tough it will likely be unless they themselves role model what it is that they are teaching. I do hope that my grandchildren will benefit from some schoolteachers who are eco-warriors or social-justice activists and who role model the changes that they teach. Please protect my grandchildren from teachers who really do not buy into the demands of sustainability but feel obliged to do their best to tell others to do so. I have more confidence, albeit still limited, that education, in general, is teaching young people to be able to think critically, and to be willing to do so, and supporting trainee schoolteachers to be able, and willing, to teach these things. Given the state of our international ESD mission at present, I do doubt that we are on the right path. I doubt that many understand the difference.

Perhaps those who teach in higher education will think, like I do, that it is not the role of higher education to teach young people what they should or should not be willing to do, other than to think critically, or (if they must) with respect to the demands of particular professions [although notably, Shephard and Egan 2018, argue that critical thinkers will decide for themselves what they will be willing to do, so additional forms of values-education may not be appropriate in any context].

Taking a U Turn

It is an irony that taking U-turns in a field of enquiry requires practitioners to not only be able to do so, but also be willing to do so. I suggest that we have three decades of misadventure to reverse and the sooner we start the better. A range of critical thinking skills and personal dispositions may be involved in such an important endeavour but from Facione (1990, section IV) and focusing on more challenging *dispositions* than on more widely available *skills*, I would recommend 'willingness to reconsider and revise views where honest reflection suggests that change is warranted' and 'honesty in facing one's own biases, prejudices, stereotypes, egocentric or socio-centric tendencies'. 'Fair mindedness in appraising reasoning' would not go amiss either. And just in case it's me, rather than everyone else, in our wonderful and diverse field of enquiry, I shall take refresher courses in these myself.

References

Bloom, B.S., et al. (1956). A Taxonomy of Educational Objectives: Handbook I The Cognitive Domain. Longman, Green Co., New York.

Bloom, B. S., Hastings, J. T., & Madaus, G. F. (1971). *Handbook on the formative and summative evaluation of student learning*. McGraw-Hill.

Brandt, J.-O., Barth, M., Merritt, E., & Hale, A. (2021). A matter of connection: The 4 Cs of learning in pre-service teacher education for sustainability. *Journal of Cleaner Production*, 279, 123749. https://doi.org/10.1016/j.jclepro.2020.123749

Brundiers, K., Barth, M., Cebrián, G., Cohen, M., Diaz, L., Doucette-Remington, S., Dripps, W., Habron, G., Harré, N., Jarchow, M., Losch, K., Michel, J., Mochizuki, Y., Rieckmann, M., Parnell, R., Walker, P., & Zint, M. (2020). Key competencies in sustainability in higher education—Toward an agreed-upon reference framework. *Sustainability Science*. https://doi.org/10.1007/s11625-020-00838-2

Evans, T. L. (2019). 'Competencies and pedagogies for sustainability education: A roadmap for sustainability studies program development in colleges and universities', Sustainability (Switzerland). https://doi.org/10.3390/su11195526

Facione, P. A. (1990). *Critical thinking: A statement of expert consensus for purposes of educational assessment and instruction*. The California Academic Press.

Facione, P. A. (2000). The disposition toward critical thinking: Its character, measurement, and relation to critical thinking skill. *Informal Logic*, *20*(1), 61–84.

Glaesser, J. (2019). Competence in educational theory and practice: A critical discussion. *Oxford Review of Education*, *45*(1), 70–85. https://doi.org/10.1080/03054985.2018.1493987

Hodge, G. (2007). The origins of competency-based training. *Australian Journal of Adult Learning 47, 2 179–209.*

Jensen, B. B., & Schnack, K. (1997). The action competence approach in environmental education. Environmental Education Research, 3, 163–178.

Krathwohl, D., Bloom, B., & Masia, B. (1964). Taxonomy of educational objectives, Handbook II: The affective domain. New York: David McKay Co.

Lakoff, G. (2010). Why it Matters How We Frame the Environment. *Environmental Communication: A Journal of Nature and Culture*, *4*(1), 70–81. https://doi.org/10.1080/17524030903529749

Norris, N. (1991). The trouble with competence. *Cambridge Journal of Education, 21*(3), 331–341.

OED (2021). Disposition, retrieved from https://www.oed.com/view/Entry/55123?redirectedFrom=disposition&

Redman, A., Wiek, A., & Barth, M. (2020). Current practice of assessing students' sustainability competencies: A review of tools. *Sustainability Science*. https://doi.org/10.1007/s11625-020-00855-1

Rieckmann, M. (2012). Future-oriented higher education: Which key competencies should be fostered through university teaching and learning? *Futures*, *44*(2), 127–135. https://doi.org/10.1016/j.futures.2011.09.005

Rodríguez Aboytes, J. G., & Barth, M. (2020). Transformative learning in the field of sustainability: A systematic literature review (1999-2019). *International Journal of Sustainability in Higher Education*, *21*(5), 993–1013. https://doi.org/10.1108/IJSHE-05-2019-0168

Sartori, G. (1970). Concept misformation in comparative politics. The American Political Science Review, 64(4), 1033–1053. https://doi.org/10.2307/1958356

Schneider, K. (2019). What Does Competence Mean? *Psychology*, *10*(14), 1938–1958. https://doi.org/10.4236/psych.2019.1014125

Shephard, K. (2008). Higher education for sustainability: Seeking affective learning outcomes. *International Journal of Sustainability in Higher Education*, *9*(1), 87–98. https://doi.org/10.1108/14676370810842201

Shephard, K. (2020). *Higher Education for Sustainability: Seeking Intellectual Independence in Aotearoa New Zealand*, Springer https://link.springer.com/book/10.1007/978-981-15-1940-6

Shephard, K, & Brown, K. (2016). How democratic is higher education for sustainable development? *Discourse: Studies in the Cultural Politics of Education*, 1–13. https://doi.org/10.1080/01596306.2016.1150254

Shephard, K., & Egan, T. (2018). Higher Education for Professional and Civic Values: A Critical Review and Analysis. *Sustainability*, *10*(12), 4442. https://doi.org/10.3390/su10124442

Shephard, K, Rieckmann, M., & Barth, M. (2019). Seeking sustainability competence and capability in the ESD and HESD literature: An international philosophical hermeneutic analysis. *Environmental Education Research*, *25*(4), 532–547. https://doi.org/10.1080/13504622.2018.1490947

Scherak, L. & Rieckmann, M. (2020). Developing ESD Competences in Higher Education Institutions—Staff Training at the University of Vechta. Sustainability, 12, 10336.

Wiek, A., Withycombe, L., & Redman, C. L. (2011). Key competencies in sustainability: A reference framework for academic program development. *Sustainability Science*, *6*(2), 203–218. 0.1007/s11625-011-0132-6

United Nations Educational, Scientific and Cultural Organization, UNESCO (2017). Education for sustainable development goals: learning objectives. UNESCO, Paris, France. https://unesd.oc.unesco.org/images/0024/002474/247444e.pdf

Complexity and Criticality in Relation to ESD Competences

7

Alain Pache and Sybille Rouiller

Abstract

This chapter first situates the paradigms of complexity and criticality in the general discourse of ESD. It then shows how the development of these two competences is absolutely central to the school. However, there is no evidence that they should be taught. Some opportunities are therefore identified, as well as some risks. Controversies about the competence approach are also mentioned, which, in the last part of the text, leads to some perspectives for teacher education.

Keywords

Complexity · Criticality · ESD · Competencies · Conspiracy theories

Introduction: Two Competences to Be Developed in the Perspective of Citizenship Education

Whatever the formulation used, the training of citizens has been one of the main aims of school for more than two centuries (Audigier 2007). Promoted by intergovernmental organisations—the Council of Europe, the European Union and the OECD—citizenship education's objective emphasises the construction of social competencies based on case studies enabling pupil to experience citizenship (ibid.). The pupil should thus be able to integrate as well as possible into society by knowing his or her rights and fulfilling his or her duties.

In order to develop such competences, the curricula recommend that 'major societal problems' such as climate change, poverty and inequality be addressed in class. The 17 UNESCO Sustainable Development Goals for schools (2017) present a more exhaustive list. These societal problems can be expressed in the form of (three-fold) controversial issues (CI). This means that these questions are debated in society, in academic discourse and in school knowledge (Legardez 2006). Working with CI thus implies dealing with uncertainties, divergences, controversies and even conflicts (ibid.).

In order to understand these CI, but also to learn how to deal with them—in particular by outlining possible solutions to sustainability problems—two forms of thinking seem particularly useful to us: complex thinking and critical thinking. Figure 7.1 schematically presents this approach, which we encourage in teacher education (pre-service and in-service). The double arrows in the diagram mean that the development of these two forms of thinking promotes the understanding and analysis of CI. But the reverse

A. Pache (✉) · S. Rouiller
University of Teacher Education State of Vaud, Lausanne, Switzerland
e-mail: alain.pache@hepl.ch

P. Vare et al. (eds.), *Competences in Education for Sustainable Development*, Sustainable Development Goals Series, https://doi.org/10.1007/978-3-030-91055-6_7

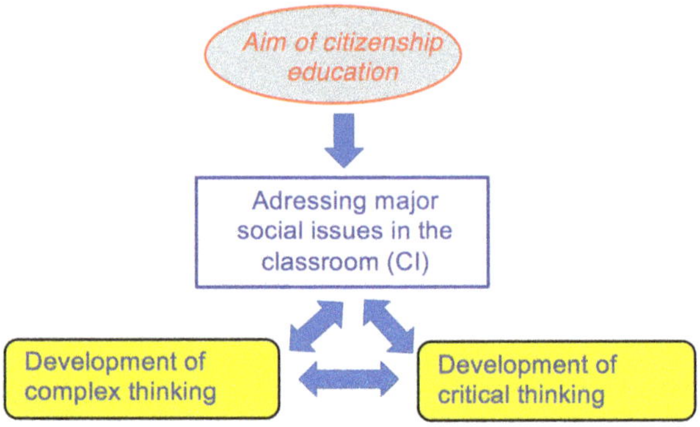

Fig. 7.1 An approach focused on complexity and criticality (based on a diagram by Ph. Hertig 2016, LirEDD)

is also true: addressing CI in the classroom promotes the development of complex and critical thinking.

The Paradigms of Complexity and Criticality

In the French-speaking world, the work of Edgar Morin (1982–1990), a French sociologist and epistemologist, has contributed to the paradigm of complexity. This is based on the application of three interrelated guiding principles (Favre 2017): the dialogical principle, the recursive principle and the hologrammatic principle. The first is defined as the complex association of instances necessary for the functioning of a phenomenon; the second emphasises the fact that in many organisations the effects interact with their cause; the third means that the whole is inscribed in each of its parts, which yet remain singular.

Morin's approach is therefore transdisciplinary, as it aims to link several elements and thus opposes the principle of duality which leads to separate and disjointed reality, which does not obey the artificial division of knowledge into disciplines that are more or less inward-looking (ibid.).

With regard to criticality, there have been many calls for the teaching of this competence in schools in many countries in recent decades. This has been particularly visible in France and Switzerland over the last 5 years with widespread media coverage of the public problem of 'conspiracy theories' and 'fake news' (Rouiller 2019).

For many authors, however, there is no consensus on the definition of criticality. According to Gagnon (2012, 2016), there are nevertheless converging elements in the various existing definitions. According to him, criticality is defined above all by its evaluative nature, which is specific to all disciplines. The evaluative mode of criticality would be directed towards five different objects (Gagnon 2016):

1. Our own conceptions or reflections (self-critical and self-correcting approach).
2. What peers have to say
3. Ideas shared by experts
4. Information from external sources (newspapers, TV, internet, textbooks)
5. The material itself (its relevance, efficiency, reliability)

Gagnon (2016) adds that the exercise of criticality must not take the form of a confrontation in which each participant in the debate tries above all to convince the opponent that he or she is right or that his or her opinion is the best. It is a matter of situating this exchange within the framework of a collaborative process of meaning in which each considers the other as capable of stating a truth in a dialogue that is open to a diversity of points of view and whose goal is to develop a vision that is as viable as possible (Gagnon 2016, p. 107). We now live in a digitised context marked by instant access to information, knowledge and discourse of very varied epistemic values. The development of criticality becomes all the more a central resource for access to intellectual and emotional autonomy and for the enlightened exercise of one's citizenship. In this sense, it is emancipatory.

Developing complex thinking and critical thinking at school (or in teacher training) can thus be understood as developing two specific competences. However, in the field of education, the notion of competence has not always been unanimously accepted. Indeed, it often provokes controversy because of the difficulty in clearly identifying the phenomena it attempts to objectify or because of methodological problems which will be developed in the rest of this text. Moreover, it questions the influence that the socio-economic context and the constraints of institutional training systems exert on the development of this knowledge (Dolz and Ollagnier 1999; Beckers et al. 2014).

The aim of this chapter is therefore to show the articulation between complex thinking and critical thinking in an emancipatory perspective of citizenship education, but also the risks and opportunities that these two forms of thinking offer in the context of education. It will therefore be a question of adopting a critical stance and questioning the idea that the competence-based approach is the most appropriate way of dealing with current social issues. We will conclude with some perspectives for teacher training.

Complex Thinking at School: Opportunities and Risks

Developing complex thinking at school involves first of all characterising it and prioritising the different tools of thought. In a seminal article, Jenni et al. (2013) first describe two components of complex thinking: first, the ability to describe and reconstruct a system in the form of an arrow diagram, which implies first identifying the elements of the system and their relationships, determining or understanding the limits of the system and then graphically representing these elements and relationships. The second step is to use the system to formulate possible courses of action on the basis of adequate arguments, which implies an ability to make predictions and evaluate courses of action.

The authors then present a model based on Morin's guiding principles outlined above, showing the 'progressive complexity' of the links forged by the students: linear relationship of cause and effect, multiple causality, positive feedback loops, recursivity loops, dialogical tension, negative feedback loops (Fig. 7.2).

The Rounder Sense of Purpose (RSP) model,[1] through the 'systems' competence, completes this work by integrating the principles of interdependence, non-linearity, self-organisation and emergence or by using different scales of analysis. The underlying components highlight specific techniques that the teacher should implement (concept mapping, systems analysis or simulation games).

Let us take the example of concept mapping. This enables the pupil or student to consider the main issues of a theme at the beginning of a teaching-learning sequence (allows representations to emerge) or at the end of learning (assessment). However, it is not always easy to achieve the desired objectives, particularly when the teacher's support is rather weak. Figure 7.3 shows a conceptual map, drawn up by students in pre-service training, which represents a phenomenon of recursivity leading to a drop in tourist activity, in other words to a development which is not really sustainable for the resort concerned. We can therefore see here that the students have been able to model an environmentally friendly dimension (or a subsystem), but without being able to really aim at sustainable development. A more pronounced underpinning would certainly make it possible to define a model taking into account all three dimensions of sustainability.

One of the limitations of the RSP model is that it does not offer a learning progression. Indeed, the different objectives and components are stated in a linear manner (1.1–1.3 / C1–C3) without any indication of where to start. This therefore requires teachers to develop a progression themselves, e.g. from simple to more complex, or by using other references (such as the one by Jenni, Varcher and Hertig mentioned above).

[1] https://aroundersenseofpurpose.eu/.

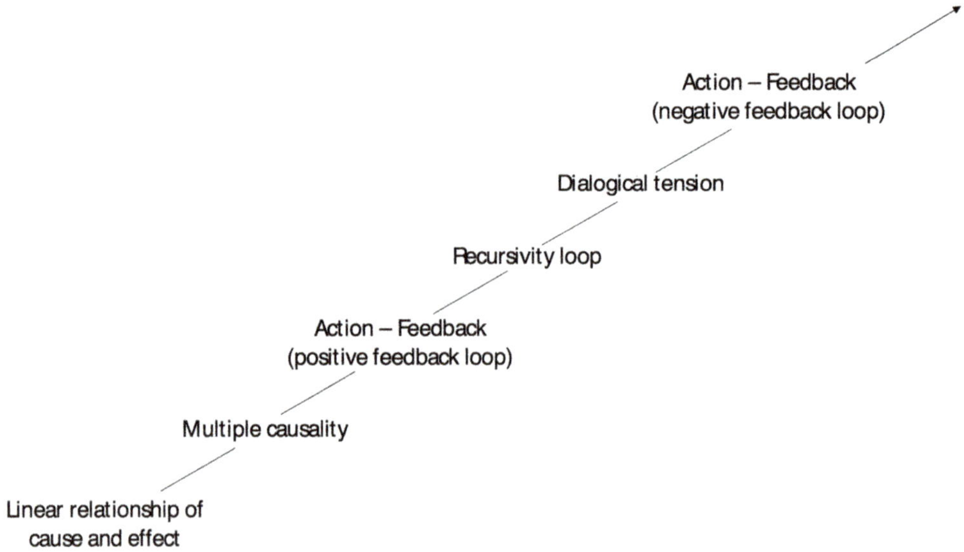

Fig. 7.2 Model depicting the progressive complexity of the links forged by pupils (Jenni, Varcher and Hertig, p. 191)

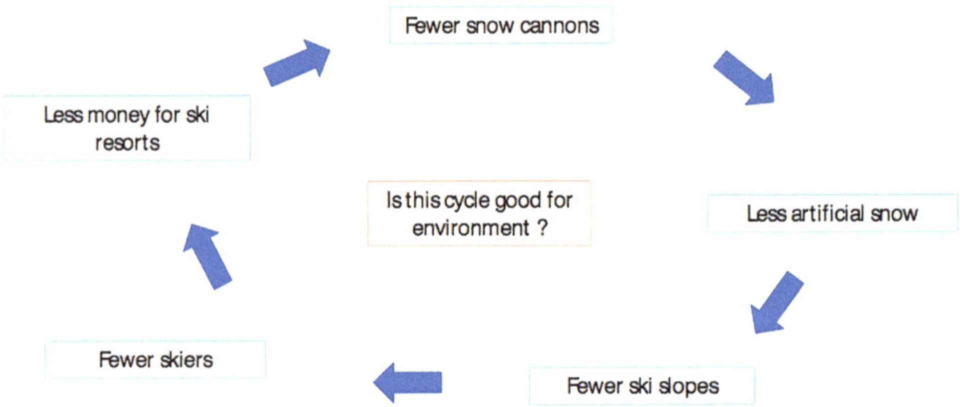

Fig. 7.3 The artificial snowmaking system (student work, Bachelor's degree, second year)

Critical Thinking at School: Opportunities and Risks

According to Curnier (2017), the starting point for critical thinking is learning to problematise, the first challenge of which is to learn to ask critical questions rather than learning the 'right' answers. Thus, critical thinking must invite learning to identify and clarify the foundations and values of a position (that of the students and those of the actors studied) (Curnier 2017, p. 208). According to him, critical thinking is essential for a way of teaching that aims at the emancipation and autonomy of citizens capable of defining solutions on an individual and collective scale to respond to the challenges of the Anthropocene. This must sometimes involve learning to think outside the frames of thought specific to Modernity and the neo-liberal socio-economic system in which we currently live and whose impacts are harmful (Curnier 2017, p. 208–209). Curnier calls on teachers to move away from the posture of transmissive learning to encourage their pupils to be actors in the construction of their knowledge. He also points out that critical thinking is based on a relationship to knowledge that favours living knowledge rather than inert knowledge (Curnier 2017, p. 210). By inert knowledge, the author means knowledge that cannot be used in action.

The RSP model of critical thinking seems at first glance to correspond to the definitions of Gagnon (2016) and Curnier (2017). However, the RSP model makes little explicit reference to disciplinary knowledge, and the examples of implementation are thus not detailed enough to be directly applicable in the classroom. For example, 'distinguishing between facts, opinions and hypotheses' (C. 4.2) requires the mobilisation of many skills but also a great deal of reference knowledge depending on the subjects covered. This can be a hindrance for teachers, especially when working on a CI. Controversial issues can present difficulties for some teachers, firstly, because they cannot rely on stabilised scientific knowledge and secondly because they are afraid that they will not be able to manage the different reactions of students to the topic. As a consequence, some teachers may decide not to treat this CI, to 'neutralise' it in order to 'cool it down' or, on the contrary, to activate it in order to 'warm it up'. Teaching choices can therefore be greatly influenced by the teacher's relationship to CI and the different ways of animating the topic (Simonneaux and Pouliot 2017).

Furthermore, according to Willingham (2007), the competence-based approach, which is dominant in the teaching and training of critical thinking, is not the most appropriate. He argues that critical thinking is not a competence that can be learned and applied in any situation. Rather, critical thinking would be a process of reflection related to the content of thought, i.e. content knowledge. Indeed, it is difficult to look at a problem from several angles without solid knowledge on the issue. A pupil can learn that they must 'think critically', but if they do not learn to put it into practice, to exercise it, and if they do not have the necessary reference knowledge, they will not be able to integrate and transfer it outside the course, because it will not make sense to them.

The Competence-Based Approach: Current Controversies and Challenges

The first works offering a critical look at competences were published, in the French-speaking world, at the end of the 1990s. In a seminal work, Dolz and Ollagnier (1999) point out that the emergence of the notion of competence in education is a sign of epistemological changes. It no longer refers to performance, but to the notion of "internal construction, the power and willingness of the individual to develop what is his or her own as a 'different' and 'autonomous' 'actor'". (p. 8). Furthermore, in the professional world, the notion of competence is used to define experience knowledge needed by people when they have to solve problems that arise in their working life.

Within the framework of school curricula, competence-oriented pedagogy defines the actions that pupils should be able to carry out after learning. This raises questions about the 'complex relationship between the logic of subject teaching and the logic of competence associated with the acquisition of complex, cross-curricular and exportable expertise outside the school field' (ibid., p. 10). Thus, depending on the case, only learning situations may be mentioned or, conversely, only knowledge may appear in textbooks. In both cases, this is unsatisfactory and does not allow for the development of competences.

These same authors denounce the ideological underpinnings of the notion of competence, in particular the idea of employability. This remark is all the more virulent today, when we look at education for sustainable development, which challenges the current model of society, and in particular the neo-liberal model.

More recently, Beckers et al. (2014) make the same argument about the risks of 'casting young people into the mould of the job structure' (p. 60). These authors also insist on the need to make schools evolve towards greater efficiency and equity, particularly by describing the types of tasks to be performed by students and the essential resources that schools should ensure that everyone is able to master. Furthermore, the aim would be to avoid the pitfall of 'invisible pedagogies' (Bernstein 1975), characterised by a weak teacher framework and offering little support to pupils when they are learning new knowledge. Numerous studies support the thesis that such pedagogies do not help pupils with learning difficulties to deal with open and high taxonomic situations (Rochex and Crinon 2011).

Another risk could be 'turning our backs on knowledge and culture' (Beckers et al. 2014, p. 64). A competence-based approach should not, in fact, lead to renouncing knowledge or exploiting it in a narrow utilitarian perspective, but to giving it 'meaning, existence and power' (ibid., p. 65). Moreover, these authors insist on the need to develop new tools and strategies to evaluate learning. This involves, for example, the teacher gathering information on several occasions and in a variety of situations. This means that when it comes to reporting on the degree of individual mastery of a competence (summative assessment), the task proposed cannot be identical to the one worked on in the learning phase. The task should therefore be new, while at the same time mobilising, in the form of new combinations, previous learning in order to respect the coherence between assessment and learning. As for formative assessment, it should be based on learning and development theories. It should be continuous and take place at all levels of teacher–pupil interaction, between pupils or on the capacity of pupils to assess themselves. Finally, these authors insist on diversifying learning targets and pedagogical approaches in order to take account of individual difficulties. Furthermore, it is important to identify the basic knowledge needed to enable all pupils to succeed. Structured and explicit teaching therefore appears to be much better suited to certain pupils than project teaching or problem-solving, provided that provision is made for situational exercises "which give meaning to the acquisition of the same prior knowledge and skills" (ibid., p. 69).

A Key Element to Address in Teacher Training: Conspiracy

One of the possible excesses that we identify in relation to critical thinking is 'conspiracy'.[2] Indeed, it is easy to imagine that by thinking in terms of criticality and complexity, our students come to adhere to theories that are not valid on a scientific level.

During her doctoral research, Rouiller observed that, paradoxically, adolescents think they are demonstrating critical thinking by adhering to certain conspiracy theories (Rouiller, in progress and Rouiller 2019). This may be accompanied by a loss of confidence and a rejection of any form of institutional discourse or epistemic authority (media, scientific, academic, etc.). Conspiracist discourses are thus identified as dissident or minority but equally valid positions as commonly accepted scientific discourses in the name of defending freedom of expression and personal opinion. Several pupils thus refer to the impossibility of 'blindly' trusting expert discourse that they cannot 'verify for themselves', thus confusing 'critical thinking' with 'excessive scepticism'.[3] For example, Leo, who spoke about the theories that cast doubt on the American moon landing in 1969:

> Léo: "Madam, I find that we can't really say if these theories are true or not if we can't see them with our own eyes. Someone can tell us but we can't know if they are telling us the truth. (...) I'm not saying I don't believe, but I can understand

[2] A distinction should be made between proven plots ('secretly concerted projects against life, someone's safety or against an institution' according to *The Petit Robert*) that punctuate history and 'conspiracy theories' that refer to plots that do not exist or that distort those that do exist (Kreis (2012, p. 7)). At a minimum, conspiracy theories would support a deep conviction that a group (a secret society, a real or fantasised minority or supernatural creatures) or an individual (often the devil or other agent of evil) is omnipotent, both elite and on the margins of society (in places inaccessible to the common person), secretly and completely controls the political and social order as well as the great historical events of which the people have only a falsified account by the media and institutions subject to this power.

[3] On epistemological individualism, see Guillon (2018).

people who say 'I don't believe'." (Léo, first year Swiss college (16 years old), in class)

In France and Switzerland,[4] a normative approach to critical thinking can be observed with regard to 'anti-compromising' educational prescriptions which, as Curnier (2017) has pointed out, are more aimed at teaching pupils 'the right answers' than at teaching them to ask themselves the 'right questions'. It is proposed, for example, to use an analytical grid in the form of a checklist that would make it possible to identify a conspiratorial discourse and thus simply disqualify it without carrying out a real critical analysis (see, for example, the French government website *'On te manipule'*).[5] This type of method takes into account neither the context in which such discourse emerges (the situation and its complex dimension, the speakers and receptors targeted) nor the knowledge that is at stake (epistemology of science, established disciplinary knowledge or unstable knowledge linked to a CI) and therefore does not develop critical thinking.

Finally, it is important to remain critical in relation to the competence-based approach. Indeed, as we have tried to show, the competence-based approach has many limitations, which makes it necessary to carry out in-depth work in the context of teacher education. In particular, it is necessary to move away from the magical way of thinking, which would consist in saying that it is enough to formulate a competence for it to be de facto acquired. This is unfortunately what we still too often encounter in our practice as trainers.

As we can see from the authors mentioned above, the competence-based approach is not a matter of course. It is demanding and requires time, significant resources and solid training (pre-service and in-service). Without this, it can be truly counterproductive.

[4]For an inventory of these prescriptions, see Rouiller (2019).

[5]https://www.gouvernement.fr/on-te-manipule accessed on 11th December 2020.

Conclusion

We have sought to show in this chapter that not all competences (such as those in the RSP model) are at the same level. In our opinion, the competences of complexity and criticality are of the highest priority, as they are prerequisites for any action in the context of sustainability. They should therefore be taught and trained initially and regularly throughout the school years. They enable a better understanding of current issues and promote emancipation. However, we have also shown in this chapter that the development of such competences is not self-evident, especially in cases where the teacher's support is weak, where the teacher does not have the necessary knowledge or where a teacher is unable to offer their pupils a relevant learning progression.

It is also important not to reify the competence-based approach, as many studies have shown that it often refers to the logic of performance (or employability) or that it neglects pupils with learning difficulties in the context of project-based teaching, for example.

Finally, we would like to stress a possible drift that would lead to adherence to conspiracy theories. This risk must absolutely be addressed in training with tools to deal with it.

References

Audigier, F. (2007). De l'usage des récits pour dire l'histoire, la géographie, l'éducation citoyenne. *Pratiques, 133–134, 77–99.*

Beckers, J., Hirtt, N. & Laveault, D. (2014). Les compétences à l'école : controverses et défis actuels. In Ch. Dierendock, E. Loarer & B. Rey (Ed.), L'évaluation des compétences en milieu scolaire et en milieu professionnel (pp. 57-70). Bruxelles : De Boeck.

Bernstein, B. (1975). *Langage et clases sociales*. Paris. Editions de Minuit.

Curnier, D. (2017). *Quel rôle pour l'école dans la transition écologique ? Esquisse d'une sociologie politique, environnementale et prospective du curriculum prescrit*. Thèse de doctorat présentée à la Faculté des géosciences et de l'environnement de l'Université de Lausanne (Suisse).

Dolz, J. & Ollagnier, E. (Ed.) (1999). *L'énigme de la compétence en éducation*. Bruxelles : De Boeck (Raisons éducatives).

Morin, E. (1982). *Science avec conscience. Points.*

Favre, D. (2017). Complexité. In A. Barthes, J.-M. Lange & N. Tutiaux-Guillon (Ed.), *Dictionnaire critique des enjeux et concepts des « Educations à »* (pp. 317-322). Paris : L'Harmattan.

Gagnon, M. (2012). Étude de la transversalité des pratiques critiques d'élèves québécois dans le cadre de leurs cours de sciences, d'histoire et d'éthique. *Revue suisse des sciences de l'éducation, 34*(3), 551-573.

Gagnon, M. (2016). Pensée critique et respect des convictions. ZFRK/RDSR, 2(2016), 103-118.

Guillon, J. (2018). Les théories du complot et le paradoxe de l'individualisme épistémique. *Diogène*, 261-262(1-2), 54-87. https://doi.org/10.3917/dio.261.0054

Hertig, Ph. (2016). *Une EDD pour une formation citoyenne.* Document non publié, LirEDD.

Jenni, Ph., Varcher, P. & Hertig, Ph. (2013). Des élèves débattent : sont-ils en mesure de penser la complexité ? *Penser l'éducation* (Hors série), 187-203.

Kreis, E (2012). *Les puissances de l'ombre. La théorie du complot dans les textes.* Paris : CNRS-BIBLIS.

Legardez, A. (2006). Enseigner des questions socialement vives. Quelques points de repères. In A. Legardez & L. Simonneaux (Ed.), *L'école à l'épreuve de l'actualité. Enseigner les questions vives* (pp. 19-31).

Rochex, J.-Y. & Crinon, J. (Ed.) (2011). *La construction des inégalités scolaires. Au cœur des pratiques et des dispositifs d'enseignement.* Rennes : Presses universitaires de Rennes.

Rouiller, S. (en cours de rédaction). Thèse de doctorat. *Les « théories du complot », l'École et les adolescents : quels enjeux, quelles prises en charge didactiques ? Analyse qualitative de pratiques et discours d'enseignants et d'élèves suisses romands et français.* Sous la direction du prof. R. Rousseleau à l'Université de Lausanne et la prof. N. Durisch Gauthier à la Haute École Pédagogique du Canton de Vaud. Université de Lausanne, Faculté de théologie et de sciences des religions.

Rouiller, S. (2019). Éclectisme et polysémie des « théories du complot » sur le Web et dans l'industrie du divertissement. Enquête ethnographique sur leur réception par des élèves (15–18 ans) suisses et français. *Studies in Religion/Sciences Religieuses*, 48(4), 593–611.

Simonneaux, L et Ch. Pouliot (Eds) (2017). Les questions socialements vives. *Sisyphus. Journal of Education.* Vol. 5 n°2.

UNESCO (2017). *L'éducation en vue des objectifs de développement durable. Objectifs d'apprentissage.*

Willingham, D. T. (2007). Critical Thinking. Why is it so hard to teach? *American Educator* (summer 2007), pp. 8-19.

8 Post-Normal Science, Transdisciplinarity, and Uncertainty in Relation to Educators' Competences: A Conversation with Silvio Funtowicz

Francesca Farioli, Silvio Funtowicz, and Michela Mayer

Abstract

The paths followed by sustainability science, post-normal science (PNS), and education for sustainable development (ESD) have crossed several times in the last 20 years. The conversation reported in this chapter highlights some of the elements that connect them and the suggestions that can be drawn for the training of educators who are aware of their role as change agents and capable of going beyond the boundaries within which education has often been confined.

Keywords

Post-normal science · Transdisciplinarity · Uncomfortable knowledge · Change agents · Education for sustainable development · Educator competences

F. Farioli (✉) · M. Mayer
IASS – Italian Association for Sustainability Science, Rome, Italy

S. Funtowicz
Centre for the Study of the Sciences and the Humanities (SVT), University of Bergen (UiB), Bergen, Norway

Introduction

Education for sustainable development (ESD) asks teachers, educators, and learners to face complex and urgent, global and local challenges characterized by a high degree of uncertainty: from climate change to the loss of biodiversity, from genetically modified organisms to global pandemics. Because we see sustainability education as a process of mutual transformative learning (we are not only educators but also learners), we constantly create a learning space to facilitate change in 'our taken-for-granted frames of reference (meaning perspectives, habits of mind, mindsets)' (Mezirow 2000, p. 7). Therefore, in the rest of the chapter 'educators' and 'learners' will be used as a shorthand (see also Chap. 11). In our work in Italy, aimed to test the 'Rounder Sense of Purpose' (RSP) competences, we have repeatedly introduced the post-normal science (PNS) approach as an important component of the social construction of knowledge, which is at the basis of Sustainability Science, together with a holistic approach and a transdisciplinary vision of knowledge.

Post-normal science, as proposed in 1993 by Funtowicz and Ravetz, affirms that present social and environmental challenges cannot be 'solved as a puzzle', simply by applying 'normal' sci-

P. Vare et al. (eds.), *Competences in Education for Sustainable Development*, Sustainable Development Goals Series, https://doi.org/10.1007/978-3-030-91055-6_8

ence, but that they need a 'post-normal' science (PNS) approach. Undeniably in 'normal' science, scientists choose the problems that, even if difficult, can be solved with the tools that science and technology provide at any given time to produce 'certain' knowledge, which is then assessed by other scientists through peer review, a quality assurance process reserved to the inner circle of their colleagues.

When confronted with policy challenges, it is traditional that decision-makers rely on purportedly objective and 'value free' science advice. However, on issues 'where facts are uncertain, values in dispute, stakes high and decisions urgent' (Funtowicz and Ravetz 1993, p. 744)—as all environmental and sustainability problems are—when problems are 'wicked', concerning different scientific disciplines and critically depending on non-scientific knowledge, scientists cannot offer 'simple solutions'. Their voice—albeit important—should be considered together with those of an 'extended peer community', where different actors bring different perspectives and different values concerning the problems at stake.

As shown in Fig. 8.1, when facts are uncertain and decisions urgent (Funtowicz and Ravetz 1993), is where the PNS approach should be applied, involving citizens in the problem-solving and decision-making processes. All the participants in the process, aware of the relevant uncertainties, can discuss not only evidence but also value differences in the search for effective actions. These elements resonate with those characterizing the RSP approach to ESD with its emphasis on values, criticality, and transdisciplinarity.

Conversation with Silvio Funtowicz

The Italian Association for Sustainability Science (IASS) had a conversation with Silvio Funtowicz aimed to clarify the connections between the PNS and the RSP vision of educators as 'change agents' with a specific and well-defined set of competences.

Silvio I am not an educator, but I think that people often fantasize on education and idealize it. The point is to understand not only what we want but what is actually achievable. Another important aspect, for all the professions that have to deal with major challenges such as sustainability, that is the focus of your project is to know how to go from here, i.e. where we are now, to there, which is where we want to arrive. We cannot afford to choose 'where to start

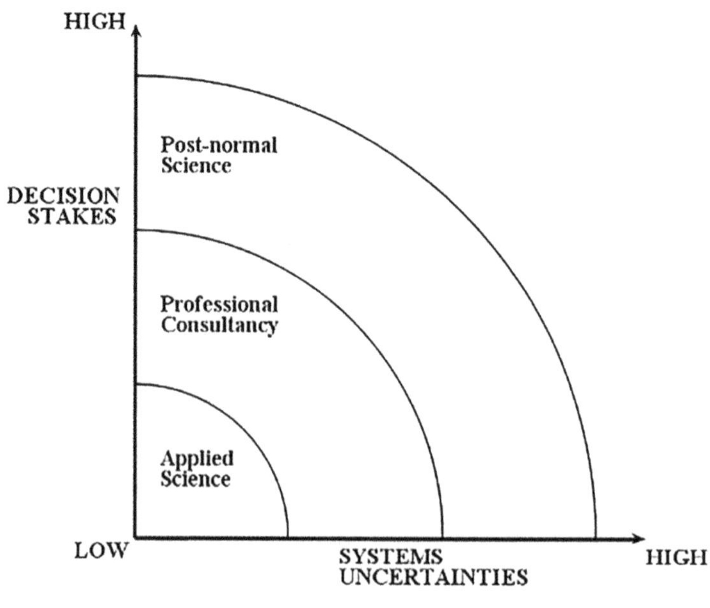

Fig. 8.1 PNS perspective (Funtowicz and Ravetz 1993)

from', we must start right from where we are now. When I taught mathematics at the University, I always asked myself this question: is there a right time for 'bad news'? When should I tell the students that the things that they had learned were not exactly how they thought they were? If you do so too early, they are not prepared, but if you wait for the moment when they are, it is often too late, and they are no longer interested.

IASS: What requests does PNS make to educators?

Silvio I can speak for myself, not for post-normal science because post-normal science is open, and everyone can have their opinion on the subject. What is important is to recognize that education takes place not only in schools or colleges, but that the young are increasingly interconnected and receiving stimuli from the entire society. The role of the educator is thus to organize these stimuli, to help the learners build their own vision, especially in times like these, when you need to cope with an increase in false or poorly founded news, constantly promoted and brought back up without shame.

We are in a different world and we need to conceive education in a different way. A first step would be to think about how we have taught science so far because science, but also the role of science, and with it that of technology and technique, are rapidly changing. And we change with them.

People believe that there is an accepted definition of science, or of scientific method, but that is not true; what we call science is not something that is established once and for all. Science has evolved, it has changed with humanity, and in my experience, academically speaking, there is a tendency to remove the memory of ideas as well as that of the scientific thought that has generated them.

IASS: If we have changed, if we are different, and circumstances and contexts are always different, what can we learn from history?

Silvio There are different ways of teaching history, just as there are different ways of teaching science. The point is starting to deal with an 'uncomfortable knowledge' (Rayner 2012), with a history and a science that are not reassuring and that show us the difficulties, the obstacles, that the changes bring within themselves. An example of uncomfortable knowledge was, in the history of mathematics, the discovery of irrational numbers by the Pythagoreans. A belief that crushed the very idea of their cosmology based on the perfection of numbers. The name 'irrational' highlights how the fears, the demons, we build for ourselves, are linked to our vision of 'normality'.

Science and society co-evolve. We can therefore mull it over and take the opportunity to consider how a scientific result is built nowadays and how it was built in the past. We can take this opportunity to teach some of the political economy of science, how 'working with science' has changed over time. Not by teaching the 'history of science' but rather by explaining how sciences are embedded in history, culture, and institutions.

Another important thing to discuss is the 'demarcation' between what science is and what it is not. If science is not only a product but also a process, how we will we be able to distinguish a scientific process from one that is not? Carnap and Popper, for instance, have tried to answer this question, but their answers were also a product of the historical context and of their beliefs about what science was not (in simple terms metaphysics and religion for the former, and psychoanalysis and historical materialism for the latter...).

It is essential to place facts in their historical context, considering not only the change of environment but also how we ourselves have changed. In the past 50 years we have changed as fast as our context, yet in predictive simulation models humans have not changed. On the contrary, we are the actors of change, as we are in regard to science and technology.

IASS: Educators should therefore be aware of being 'agents of change'. But how can they use the post-normal vision of science and address the discomfort that inevitably comes from suggesting an 'uncomfortable knowledge'? School and society introduce an image of human evolution, and of its recent history, as a path of continuous progress. It is still done in school to introduce the Darwinian theory of evolution: there have been dark moments and moments of glory, but progress towards 'modernity' is assured (Gould 1990). How can we guarantee a critical vision not only of the history of science but also of the history of humanity, and therefore of present-day modernity?

Silvio I do not know how, but I believe it can be done. We can begin to address this dogmatic view of what science is. Alternative narratives can be presented, stories that can explain, somehow better, what has happened. But naturally a lot has been invested in the way normal science is taught, as well as other subjects, and it is difficult to change. What I see is that even among the front-line scientists – especially those who deal with biodiversity, climate change, but also with epidemics, with COVID-19 – the crisis is very deep, it is an existential crisis. And it is on the crises that we must act, we must reassure ourselves that wanting to change things is perfectly normal, it is *not* a disease.

We must establish the idea – with learners and with educators – that doubt and scepticism are not a pathology, but a natural state of the human condition.

The problem is a practical one; it is about how to do it. At this very time, *what* is done is not important, as opposed to *how* it is done. Products are the result of processes, and if the processes are of good quality, the products will be too, and they will be able to create change.

It is essential to abandon the idea of 'control' – a Cartesian and Laplacian heritage – and the idea, the illusion, that there is an established blueprint when in reality there is none. We have a problem of legitimacy and credibility in explaining that we do not control; quite the opposite of we have been told for a long time, that there are no established solutions, and perhaps there is no solution at all.

We need to know that it is necessary to experiment and accept the possibility of failing, and that failure is not a tragedy. The real tragedy is believing that we know where we are going.

IASS: Dealing with uncertainty is a key competence for our RSP proposal, which is heading towards the PNS and so is the importance of building participation and experimenting through transdisciplinarity, giving voice not only to science but to those who experience and work directly on problems.

Silvio Yes, but these competences, as you have expressed them, do not have a unique meaning, and I can, as an educator, act without changing what we are doing. Because change is difficult, and too much has been invested in continuing to do more or less what we did before. The very same Sustainable Development Goals (SDGs) offer indicators that continue to recreate the same type of data and analysis tools, those that allow us to say, 'we are approaching the goal'. Whereas the real challenge is to say that everything needs to be changed, and to be optimistic and think that it can be done.

PNS argues that a great diversity is necessary, coming from the bottom, and involving great changes, even in the way of thinking. We are living in an age of transition, like the one following the invention of the printing press: a new and disruptive technology that helped to question existing power relations and the monopoly of culture. We are also living in a period of extraordinary change and as with the Sixteenth Century, changes do not necessarily take place through the existing institutions, such as university departments or research establishments, where what is studied increasingly depends on the available funding.

IASS: The role of educators should be precisely that of opening everyone's mind to possible worlds, but how can we build different ways of thinking with young people or people in general who are totally immersed in this world, in a reality that is thought of as the only possible one?

Silvio And that is exactly where we need to start from, from young people who communicate by exchanging text messages in their smartphones but who are also capable of asking troublesome questions. And COVID-19 perhaps offers us an opportunity because it is easier to reflect on the meaning of normality in a context of emergency.

IASS: We like to imagine the educator's role to be a transformative one. Can your words provide a starting point towards a necessary transformation?

Silvio We can ask ourselves about the meaning we give to words: when speaking of participation what do we mean? In my opinion it is very clear, there is participation if there is 'partnership' and hence educators cannot separate their interventions from other moments regarding relationships and formation: peers, parents, friends, associations. I am particularly interested in transdisciplinarity, and how to embed scientific knowledge in a broader concept of knowledge. With COVID-19, people are discovering that 'knowledge' does not speak with just one voice but with a multiplicity of voices. Another thing worth teaching is practical knowledge, rediscovering, for instance, that science is also a craft. Many problems in history have been solved by 'practical knowledge', from that concerning the measure of longitude to that of building and maintaining Brunelleschi's Dome in Florence. Equations were not enough, there was a need for practical knowledge, of which the artisans of that time were producers and bearers.

Fundamental is to know *how* and not only to know *that* ... Understanding how other civilizations have managed to solve tangible and urgent problems, even without the support of science as we understand it today. The risk is that when we talk about 'knowledge', it all comes down to 'science' or technology. For example, Covid-19 vaccines and fundamental research are very important, but practical measures like wearing a mask, washing hands and distancing are things we have known for years, things that come from experience, and that sometimes – as in the case of Semmelweis[1] – have even been in contrast with the science of the time.

I think that a good way of introducing PNS would be to 're-categorize knowledge', beginning from that which has to do with experience, with living together, with the body, rediscovering a 'feminist epistemology'. The idea of a 'strong objectivity', contrasting the mainstream perspective of science as a 'view from nowhere' (Nagel 1986; Harding 2005). The PNS provides a vision that comes from 'many parts', in its plurality, recognizing the visions of others, not to reach a consensus but mostly to learn how to work in diversity.

All the competences you have proposed are fine, but let us strive to give them an operational aspect: what do they basically mean? How do we propose things that are different to the ones that no longer work? How can we recognize what lies beyond science, formalized disciplines, academies and understand how we can use what is out there as a resource and not as a problem to be caged in existing structures? This is the starting point for increasingly transformative considerations and actions.

IASS: But how can we be positive, optimistic, even when we are presented with an 'uncomfortable knowledge'?

Silvio This is partly a problem of those generations born after the Second World War who believe that all rights are acquired, while in reality each generation should fight for its own rights. We belong to a very particular moment in the history of humanity: the rest of our history is way more tragic. It is mostly our responsibility, as we have tried to deny the existence of this 'uncomfortable knowledge', obviously to protect our children, but in doing so we have also deprived

[1] Semmelweis was a doctor who in the nineteenth century discovered the causes of puerperal fever and proposed basic rules of hygiene to be followed, such as washing hands, so to avoid infection. Despite the impressive results when applied, Semmelweis's ideas were rejected by the medical community of the time who felt offended by the suggestion that their hands were unclean and found his theory of 'cadaverous particles' too far from their theories of disease to credit. As a consequence, obstetric practices did not change, and puerperal fever continued to be a major cause of maternal mortality (Weatherall and O'Connor 2020).

them of certain competences. And we continue to do so when we try to solve their issues instead of making them deal with them on their own.

Final Reflective Thoughts

Our conversation allowed us to identify some of the challenges that the PNS shares with ESD, but also some aspects that, as educators, we should expand.

Many of the competences for educators that have been suggested in recent years (see Chap. 3) and many of the pedagogical approaches already in use (see Chap. 17) can be re-examined in light of PNS. However, this reflection on PNS itself allows us to emphasize some aspects and suggest a few warnings:

- In the first place, the importance of a co-creation of knowledge: Hence not only bringing together different disciplines and not only by enhancing both scientific and practical knowledge in society, but also by recognizing the contribution that different visions of the world, different 'standpoints', can offer to help us explore possible resolutions to the wicked problems that contemporary life persistently presents us with. Transdisciplinarity, so difficult to practice in a context of formal education, involves a preparedness to listen to every voice while simultaneously teaching us to become accustomed to checking evidence, also to recognize the strategic use of false information, so easily amplified by information and communication technologies, and to accept, if not welcome, the 'uncomfortable' presence of legitimate discordant voices.
- There is a need therefore to dwell increasingly on the notion of uncertainty, something that should always be accompanied by that of democracy and by that of a democratic society where 'no problem is solved in advance' (Bauman 2000, p.90)[2] and 'uncertainty does not cease once a solution is adopted'[3] (Bauman 2000, p.90).
- In this approach, errors or ignorance should no longer be seen as problems to be avoided but instead, as lessons to be learned and on which, even partially, we base our decisions:

> The philosophical perspective ... is one of the complementarities of 'knowing- that 'and 'knowing-how', where uncertainty and quality are essential attributes of knowledge… where there is a dialectic interaction of knowledge and ignorance. In this way ignorance is usable, indeed useful, and essential, for the understanding of ourselves and our relations with our environment. Awareness of ignorance can be the beginning of wisdom about our place in the contemporary world… [4] (Ravetz 1992, p. 101).

- Finally, both ESD and PNS, require a different kind of education, where *how* to teach, i.e. the quality of educational processes, has the same, if not more, importance as *what* to teach. But what do we mean by the 'quality' of an educational process, and how can we recognize it when looking into the competences of educators? The RSP project has attempted to define quality by proposing learning objectives, associated with competences, a set of 'underpinning components', examples of observable behaviours, which would provide 'clues' on the quality of the processes that have been used. During the course of IASS' work in the Italian context, we have introduced and used them both for self-assessment as well as in action-research processes. But further research is needed.

The quality of processes, whether they are educational or not, depends on many factors and, as suggested by post-normal science, it must be contextualized and shared. There is no single point of view, but it is the comparison with other points of view that allows us to expand, and at the same time analyse in depth what this quality is that we are talking about.

[2]Translated by authors.

[3]Translated by authors.

[4]Translated by authors.

References

Bauman, Z. (2000). La solitudine del cittadino globale. Feltrinelli Editore, Milano. Original work: Bauman Z (1999) *In search of politics* (trans: Bettini G). Stanford University Press

Funtowicz, S., Ravetz, J. (1993). Science for the post-normal age. *Futures*, 31(7): 735-755

Gould, S. J. (1990). Wonderful life: The Burgess Shale and the nature of history. WW Norton & Company

Harding, S. (2005). Rethinking Standpoint Epistemology: What is Strong Objectivity? In: Cudd A, O. Andreasen R (ed) *Feminist Theory: A Philosophical Anthology*. Wiley-Blackwell Publishing, Oxford

Mezirow, J. (2000). Learning to think like an adult. In: Mezirow J et. al (eds) *Learning as transformation: Critical perspectives on a theory in progress*. Jossey-Bass, San Francisco, p 3-33

Nagel, T. (1986). *The View from Nowhere*. Oxford University Press, Oxford

Ravetz, J. (1992). Connaissance utile, ignorance utile? In: Theys J, Kalaora B (eds.) *La terre outragée. Les experts sont formel!* Autrement: Paris, 87-103

Rayner, S. (2012). Uncomfortable knowledge: the social construction of ignorance in science and environmental policy discourses. *Economy and Society* 41(1): 107-125

Weatherall, J.O., O'Connor, C. (2020). Conformity in scientific networks. *Synthese*. https://doi.org/10.1007/s11229-019-02520-2

9 ESD Competences for Deep Quality in Education

Chrysanthi Kadji-Beltran and Aravella Zachariou

Abstract

'Ensure inclusive and equitable quality education and promote lifelong learning opportunities for all' is the fourth of the 17 Sustainable Development Goals. In practice, education for sustainable development (ESD) provides a well-defined context for quality education and competence-based ESD that can potentially become a framework for reaching effective and deep quality education if certain conditions are met. In this chapter we elaborate and reflect on the concept of quality education with respect to different education models and re-visit quality, not only in terms of measurable, quantitative outcomes but qualitative, subtle elements of quality in education. These elements can be sought within the context of education, the competences it is expected to deliver, appropriate pedagogies, its effectiveness and relevance and most importantly reflexivity and sustainability. Integrated in a competence-based ESD, 'qualitative elements' can ensure a deeper form of quality aligned to the humanistic education paradigm and values-based critical pedagogy. In this context we critically scrutinize an international model for competence-based ESD and discuss how such models can contribute to quality education.

Keywords

Education for sustainable development (ESD) · ESD competences · Competence models · Quality education · Deep quality education

Introduction

Quality Education (QE) finds its way into educational systems through the well-defined context, solid knowledge and pedagogy of education for sustainable development (ESD) as they share a common educational vision and objectives (Kadji-Beltran et al. 2017) in terms of pedagogy, skills and competences. QE can also benefit from the connection that ESD has with competence-based education, which enables the transition towards outcome-oriented and effective education. In this case, elements of quality must be defined and addressed within a competence-based ESD as competence-based education is often criticized for quantifying and standardizing education in alignment with an economic model of education instead of ensuring real quality. Following this rationale, the current chapter presents education for sustainable development as

C. Kadji-Beltran (✉)
Frederick University, Nicosia, Cyprus
e-mail: pre.kch@frederick.ac.cy

A. Zachariou
Frederick University, Unit of EESD/CPI, Nicosia, Cyprus

P. Vare et al. (eds.), *Competences in Education for Sustainable Development*, Sustainable Development Goals Series, https://doi.org/10.1007/978-3-030-91055-6_9

QE and focuses the discussion on whether and how competence-based ESD enables a more effective form and delivery of QE, thereby promoting the achievement of the fourth Sustainable Development Goal.

Quality Education

In an attempt to define quality in the context of education, Colby, Witt and Associates (2000) claim that there is a consensus on the following dimensions of QE: good health and nourishment that allow the participation of the learner; gender sensitive, safe and well-equipped learning environments; curriculum content and materials that ensure the acquisition of basic skills and knowledge in areas such as literacy, numeracy and life-skills; as well as gender, health, nutrition, HIV/AIDS prevention and peace. Quality is also reflected in the child-centred pedagogy and assessment used in well-managed classrooms and 'nourishing' school environments resulting in reduced disparities. Quality outcomes encompass knowledge, competences, skills and attitudes linked to national educational goals and a positive contribution to society. Teacher-education can support quality by placing special emphasis on teachers' pedagogical content knowledge, pedagogical knowledge and content knowledge (Evens et al. 2018).

Nikel and Lowe (2010), synthesizing many studies on QE, propose a framework of seven dimensions of quality that are held in dynamic tension: (a) Effectiveness: the extent to which stated educational aims are met; (b) Efficiency: economic considerations, such as ratio of outputs to inputs to maximize the use of resources; (c) Equity: issues of access to education for all people regardless of gender, ethnicity, disability, sexual orientation, etc.; (d) Responsiveness: meeting the needs of the individual learners in classroom interactions by taking into consideration the uniqueness of the learner's abilities; (e) Relevance: the usefulness of education to the life of the learner immediately, when the learner comes of age and to a more distant future later in the learner's life; (f) Reflexivity: the ability to adjust to change, especially rapid change, which is important in engaging with an uncertain future; (g) Sustainability: goal-setting, decision-making and evaluation that attends to 'the longer-term future over the present and to the global as much as the local' (p. 599).

Providing QE is a challenging undertaking partially because the concept of quality in education is continuously evolving. Education is a complex system embedded in a political, cultural and economic context. The goals and orientation of education are influenced and defined by each country's contemporary reality and norms in a local and global perspective determined by their historical-cultural, anthropologic, biophysical and sociocultural horizons (Xohelis 2018). This reality is infused in the programmes of study, shapes their cognitive, epistemological, technological and scientific orientations (Herrerias-Lopez 2010) and influences the perceptions of quality which may be pursued in the form of efficiency, effectiveness and equity (Colby, Witt and Associates 2000).

International literature highlights three predominant models in education that influence how quality is perceived (Laurie et al. 2016):

1. The *economic model* of education deals with inputs and outputs. It represents an 'economist' view of education using quantitative and measurable outputs to measure quality. Quality is connected to enrolment, retention and drop-out numbers, rates of return on investment in education in terms of earnings and measurable students' achievement through standardized national or international tests. Within this theory, education is a key factor to economic development and reduction of poverty. Performance measures connected to quality entail standardized tests, certification, student satisfaction measures, industry feedback, international quantitative measures, national indices, audits to set standards and student evaluations.
2. The *humanist paradigm* emphasizes education as a process, focusing on an empowering, holistic development of the students' personality (Xohelis 2018). The humanist tradition is

based on the observation that children have an innate interest and ability to learn. They achieve personal goals, such as development of personal talents and abilities and wider social goals, e.g. understand cultural elements, respect human rights and act in ways that ensure social justice and democracy. Humanist approaches are currently described with terms such as learner-centred, participative and democratic and aim to develop creativity and problem-solving abilities amongst other goals. UNESCO (2005) has also identified common ground in the discourse around quality to include 'respect for individual rights, improved equity of access and of learning outcomes, and increased relevance which align quality to the humanist tradition' (p.19).
3. A third model of QE '*learning as connection*' emerged from sub-Saharan Africa. Following a constructivist perspective, the researchers that elaborated this model argue that for QE in the African context it is important to address issues that threaten sustainability and bring everyday knowledge into relationship with abstract and academic concepts so that both can grow together (Lotz-Sisitka 2013 in Laurie et al. 2016). This model is very similar to the humanistic paradigm, but it proposes a specific methodology of implementation.

The quality dimensions identified by Colby, Witt and Associates (2000) and Nikel and Lowe (2010) are prerequisites for reaching the shared vision of the Education 2030 Agenda which reflects a commitment to helping learners to develop as a whole person, fulfil their potential and participate in building a common future that ensures the well-being of individuals, communities and the planet (OECD 2018). In order to reach this envisioned future we need to overcome environmental, economic and social challenges. Humanistic education can help learners abandon the notion that resources are unlimited, and instead value common prosperity and well-being and develop the competences needed in order to be active, responsible and engaged citizens. Therefore, the purpose of education goes far beyond preparing young people for the world of work and entails a 'deeper' definition of quality. If we are to address quality as education for sustainable development we have to explore and define the aspects of 'Deep Quality' in that context.

Sustainable Development Goal 4 aims to 'Ensure inclusive and equitable quality education and promote lifelong learning opportunities for all', in doing so it sets seven targets addressing aspects of quality that refer to equitable access to QE and relevant and effective learning. Special reference is made to preschool education and technical, vocational and university education. Targets a, b and c concern the school climate and infrastructure, financial support of students especially from the least developed countries and preparation of qualified teachers to deliver QE. Each target's corresponding indicator measures quality in terms of participation rate, percentage or proportion in which the target is achieved emphasizing quantifiable measures and aspects of quality, except for the target 4.7 UN (n.d.).

Although the quantifiable elements of quality are important, in the context of ESD it is imperative to stress the qualitative elements of QE that would result in a stronger form of ESD and provide the education needed for the 2030 horizon. These can be sought within: the context of education (cognitive goals, emotional goals and skills); the competences that education is expected to deliver; appropriate child-centred pedagogy; the effectiveness of education; responsiveness and relevance in meeting everyone's needs and, perhaps most importantly, reflexivity and sustainability. Integrated in a competence-based ESD, qualitative elements can ensure a 'deeper form of quality' aligned to the humanistic education paradigm and values-based critical pedagogy (Herrerias-Lopez 2010).

ESD, Competences and Quality Education

A competence-based education seeks to ensure efficacy, effectiveness and efficiency since it is output-oriented (Adomβent and Hoffman 2013).

For ESD as well as QE, competences are particularly important as they can bridge the gap between knowledge and action (Rieckmann 2018) and respond to both the 'humanistic' and the 'learning as connection' models of quality.

Competence-based education is linked to a focus on improving employability (Rieckmann and Hericks 2016 also see Chap. 2) and as such it has been criticized for falling under the economic model of education which undermines how QE is interpreted and is linked to a superficial form of ESD. Similarly, the discussion around QE pivots towards a 'version of market-driven education' (Gutiérrez 2016, p. 187) and our inability to imagine new trajectories for interpreting and acting in the world we are part of. Our educational vision is currently unable to transfer transformative experiences necessary for encouraging learners to act with agency (Jickling and Sterling 2017). Transformative competences are imperative for responding to the complex fluidity in today's world and the tensions generated.

Ensuring quality within a competence-based education is a challenge as the quality of the competences determines the quality of the programmes developed. Different models and innovations emerge rapidly although the process of improvement and effectiveness is long-term and iterating along the way (Sturgis and Abel 2017).

Bral and Cunningham (2016) explored and identified nine shared elements of quality within competence-based educational programmes: clear-cross cutting and specialized competences; coherent competence-driven programme and curriculum design; embedded process for continuous improvement; enabled and aligned business processes and systems; engaged faculty and external partners; flexible staffing roles and structures; learners' orientation; measurable and meaningful assessment; new or adjusted financial models.

Seeking quality in competence-based education (CBE), Sturgis and Abel (2017) discuss four important components. The first component places 'quality' at the core of CBE structure (beliefs, policies and operational processes) through deconstructing the traditional structure and constructing a new one with great intentionality to ensure that it works effectively. The domains that need to be restructured include, among others: the mission, a transparent continuum of learning objectives, students' performance, growth and process, mechanisms of empowerment for teachers and students, school autonomy and flexibility to provide timely differentiated support to students, internal accountability, evidence-based professional learning for teachers and adaptive leadership and empowering management. The second component focuses on quality learning, safety and respect, reflected in all elements of school culture. The third component of quality is identified in the pedagogy used; a clear philosophy of learning is a critical element of quality structure and facilitates the transition towards CBE. The fourth component addresses learning experiences; educational structures and mechanisms are needed to support quality learning experiences which can have a transformative effect.

The OECD (2018) document: *The Future of Education and skills, Education 2030* identifies three categories of 'transformative' competences to transform our society and shape our future, that address the growing need for innovative, responsible and aware learners and bridge the gap of a market-oriented quality and competence-based education (pp. 5–6):

(a) The 'creating new value' category entails creative thinking for new jobs, products, processes and methods, new ways of thinking and living, business and social models. Innovation springs through collaboration with others and requires adaptability, creativity, curiosity and open-mindedness.
(b) The category of 'reconciling tensions and dilemmas' is imperative in a world of interdependency and conflict. People will be able to ensure their individual and collective well-being by understanding the needs and desires of others and reconciling any tensions. This requires a more integrated and systemic way of thinking.

(c) 'Taking responsibility' is a prerequisite of both reconciling tensions and dilemmas as well as creating new value. It suggests a sense of responsibility and moral and intellectual maturity with which a person can reflect upon and evaluate their actions with respect to their experiences, personal and societal goals and what is morally right or wrong.

Jickling and Sterling (2017) argue that education needs to be 'remade in ways that are conceptually strong and respond to the educational imperatives of our time, particularly as they relate to ecological crises and human/nature relationships' (p. 1) through deconstructing the old and reconstructing a new vision for education oriented towards ESD. Nevertheless, ESD can become an empty signifier (p. 4) if it is not outcome-oriented and closely connected to QE.

So, how can we prevent ESD from becoming an empty signifier and at the same time ensure that it can support a 'deeper' form of quality in education? Can ESD competences guarantee this synergy?

The RSP Model of ESD Competences and Deep Quality in Education

In order to further reflect on how ESD competences can support Quality in Education we map qualitative elements of QE resulting from our literature review with Competence-Based Education (Table 9.1). We focus on A Rounder Sense of Purpose (RSP) as a model not determined by national or social factors (O'Flaherty and Beal 2018), but one addressing educational institutions as communities where transformation can be achieved through an iterative learning loop (Vare et al. 2019, p. 9).

The RSP model has the potential to ensure quality in terms of the context of education as it is flexible and dynamic. It can be adapted to a variety of educational frameworks. The competences, as presented on the RSP website (https://aroundersenseofpurpose.eu), are simple, visibly interconnected and linked to the 17 UN Sustainable Development Goals (SDGs). They give educators and learners a sense of purpose and empower them to meet the SDGs in pursuit of a better life for all. Having the RSP ESD

Table 9.1 Mapping elements of quality in competence-based education with the RSP model

Quality in education	Quality in competence-based education	The RSP model
Context of education	Quality learning	Interdisciplinary Linked to 17 SDGs
Competences	Specialized competences Quality learning	12 ESD competences
Pedagogy	Appropriate pedagogy Child-centred pedagogy Quality learning	Child-centred educational approaches, collaborative learning, suitable educational strategies and techniques (concept mapping, debate, role play, modelling, investigations, project-based learning, etc.)
Classroom/school environment	Quality at the core (beliefs, policies, operational processes) Quality learning	Whole institution approach
Effectiveness	Processes for continuous improvement Measurable and meaningful assessment Quality learning	Learning outcomes and underpinning components for each competence
Responsiveness	Learning experiences	Flexible to meet different learning needs
Relevance	Quality learning	Real-life issues—Connected to raising quality of life for all
Reflexivity	Quality learning	Competences empower learners to be flexible, adaptable and pro-active to address present and future challenges

competences as a guide, the SDGs can be integrated in each educational system as real-life issues, specific in the particular social, cultural and political context, giving relevance to learning. The RSP model can also ensure quality in education in terms of the content of education. Its connection to Content Knowledge and Pedagogical Content Knowledge is reflected on what both the educator and the learner need to know.

Evolving from more abstract and complex constructs of ESD competences, RSP distils and defines 12 competences (a manageable number) using accessible and user-friendly language. Each competence and its related learning outcomes are explicitly explained and analysed through the underpinning components provided in each case. This provides guidance as to what has to be achieved and how progress can be monitored by evaluating outcomes and ensuring effectiveness. The competences in the RSP model are not operated as another 'laundry list' (Wiek et al. 2011), but seek to contribute to QE by empowering educators to surpass school effectiveness and performance, ranking and quantification and reach meaningful school development. The model includes elements that go beyond the instrumental character of education; through the RSP lens educators address schools as whole institutions. It enables multiple roles for the educators and suggests the reform of learning itself based on re-thinking, re-discovering and re-designing. This is evident through the activities proposed by the model's webpage, which are adjustable to the level of the classroom and to what is being taught and learned according to the curriculum as well as proposing the use of school grounds, integrating school policies and seeking school collaboration and networking with local agents in the learning experience of ESD. These attributes of the RSP model and its related activities can help teachers increase the relevance and effectiveness of their teaching.

Pedagogy is particularly important in QE as it is the essential element in the educational process that enables teachers and learners to critically examine the world in which they live, identify problems and find viable solutions (UNESCO 2005). The RSP model places special emphasis on the pedagogy of the competences. The webpage of the project hosts a rich pool of activities, approaches and strategies. The proposed activities go beyond traditional ESD approaches and address all of the '5P' principles of the SDGs: planet, prosperity, people, partnership and peace, aiming to shape citizens that will be more inclusive in their perceptions of the world, able to differentiate its various aspects, open to other points of view and able to integrate differing dimensions of their experience into meaningful and holistic relationships.

Finally, the RSP model holds an important element of quality, the element of reflexivity. Learners become aware of their own role and moral responsibility and are able to adjust to change through developing the competences for envisioning a sustainable future and acting towards its achievement.

Concluding Remarks

This chapter possibly represents one of the first attempts to link QE to ESD competences acknowledging that these are of paramount importance in reforming educational systems despite their complexity in context and interconnectedness. Their discussion is timely, as what is currently required by education is quality in opposition to instrumentalism and mere quantification. The concern about quality is reflected in the various ESD competence models that have been developed (Corres et al. 2020; UNECE 2012; Vare et al. 2019). Despite being flexible and accessible, the RSP model has a strong theoretical character, while its flexibility in terms of evaluation has not been addressed yet. There is a need for developing appropriate tools that can evaluate educators' adequacy and effectiveness in using ESD competences to achieve quality in education (see Sect. 3 of this book).

The RSP model is not the only model that can promote QE. Quality is intrinsically pursued by most models. What needs to be defined is *which quality* and with *which terms*. Without claiming that we answer these questions, we hope that we have triggered reflection and dialogue with respect to QE on the basis of its context and nature (qualitative elements and elements of quality in QE) as the key for a clear definition, interpretation, understanding and implementation of competences in ESD.

References

Adomβent M, Hoffman T (2013) The concept of competencies in the context of education for Sustainable Development. https://pdfs.semanticscholar.org/9ec3/118c915b2b11fd1017a1691398346f46af45.pdf

Bral C, Cunningham J (2016) Foundations of quality in competency-Based programs: Competencies and assessments. The Journal of Competency-based Education, 1:118– 121. https://doi.org/10.1002/cbe2.1027

Corres A, Rieckmann M, Espasa A, Ruiz-Mallén I (2020) Educator Competences in Sustainability Education: A Systematic Review of Frameworks. In: Sustainability 12(23), 9858. https://doi.org/10.3390/su12239858

Colby J, Witt M, and Associates (2000) Defining Quality in Education, UNICEF.

Evens M, Elen J, Larmuseau C, Depaepe F (2018) Promoting the development of teacher professional knowledge: Integrating content and pedagogy in teacher education. Teaching and Teacher Education, 75: 244-258. ISSN 0742-051X.

Gutiérrez, KD (2016) Designing resilient ecologies: Social design experiments and a new social imagination. Educational Researcher, 45(3), 187–196

Herrerias-Lopez, JA (2010) Education for Quality through Competences for competitiveness or cooperation [Educación de calidad y en competencias para la competitividad o para la cooperación]. Revista Compultense de Educación 21(1): 107–122

Jickling B, Sterling S (Eds) (2017) Post-Sustainability in Environmental Education: Remaking Education for the Future. Palgrave.

Kadji-Beltran C, Christodoulou N, Zachariou A, Lindemann-Matthies P, Barker S, Kadis C (2017) An ESD pathway to quality education in the Cyprus Primary Education Context. Environmental Education Research. DOI: https://doi.org/10.1080/13504622.2016.1249459. 23(7): 1015-1031

Laurie R, Nonoyama Y, McKeown R, Hopkins C (2016) Contributions of Education for Sustainable Development (ESD) to Quality Education: A Synthesis of Research. Research. 10(2): *226–242* https://doi.org/10.1177/0973408216661442 Centre for Environment Education, Ahmedabad, Gujarat. India.

Lotz-Sisitka H. (2013). Conceptions of quality and 'learning as connection': Teaching for relevance. Southern African Journal of Environmental Education 29:25–38.

Nikel J, Lowe J (2010) Talking of fabric: A multidimensional model of quality in education. Compare 40(5):589–605

OECD (2018) The future of Education and Skills. Education 2030. OECD.

O'Flaherty J, Beal E (2018) Core competencies and high leverage practices of the beginning teacher: a synthesis of the literature. Journal of Education for Teaching: International Research and Pedagogy 44(2):1-18

Rieckmann M. and Hericks N. (2016). Competencebased Education and its Contributions to Quality Enhancement in Higher Education. Conference: International Consortium for Educational Development (ICED) and HELTASA Joint Conference. Cape Town, South Africa.

Rieckmann M (2018) Learning to transform the world: key competences in Education for Sustainable Development. In Leicht, A. Heiss J. and Byun W.J. (eds), Issues and Trends in Education for Sustainable Development. UNESCO, Paris.

Sturgis C, Abel N (2017) In search of Efficacy: Defining Elements of Quality in a Competency-based Education System. Content in this report is licensed under a Creative Commons Attribution 4.0 International license. iNACOL.

UN (n.d.) SDG4 Ensure inclusive and Equitable Quality Education. https://sdgs.un.org/goals/goal4

United Nations Economic Commission for Europe UNECE (2012) Learning for the future: Competences in Education for Sustainable Development, ECE/CEP/ AC.13/2011/6. https://www.unece.org/fileadmin/DAM/env/esd/ESD_Publications/Competences_Publication.pdf

UNESCO (2005) Global Monitoring Report 2005: Education for All – The Quality Imperative. UNESCO

Vare P, Arro G, de Hamer A, Del Gobbo J, de Vries G, Farioli F, Kadji-Beltran C, Kangur M, Mayer M, Nijdam C, Millican R, Réti M, Zachariou A (2019) Devising a competence-based training programme for educators of sustainable development: lessons learned. Sustainability 11(7), 1890; doi:https://doi.org/10.3390/su11071890.

Xohelis P (2018) Introduction in Pedagogy. Fundamental Issues of Pedagogical Sciences [Εισαγωγή στην Παιδαγωγική. Θεμελιώδη προβλήματα της Παιδαγωγικής Επιστήμης]. Kyriakides publications. Greece.

Wiek A, Withycombe L, Redman CL (2011) Key competencies in sustainability: a reference framework for academic program development. Sustainability Science 6: 203 – 218. https://doi.org/https://doi.org/10.1007/s11625-011-0132-6

Part II

Integrating Competences

The Highs and Lows of Working ESD Competences into HEI Frameworks

10

Isabel Ruiz-Mallén, María Heras, Ramon Ribera-Fumaz, Hug March, and Andrea Corres

Abstract

Embracing the UN Agenda 2030 is a challenge for Higher Education Institutions, especially for those that already have well-established competence frameworks but without extensive experience in sustainability education. Among other things, they must embed sustainability values in their strategic academic planning and organizational work. Building capacities among their teaching staff is one of the first steps to achieving these goals. This chapter presents a case study of an online training course organized by the Universitat Oberta de Catalunya (Barcelona, Spain) that had the twofold objective of helping teachers address the UN Sustainable Development Goals in their teaching and improving their ethical competence. The chapter closes with reflections on the difficulties and opportunities of integrating ESD competences into pre-structured institutional models.

Keywords

Competence-based approach · Ethical competence · Gender · Institutional framework · Online training

Introduction

In 2015, the UN launched the Agenda 2030 with its 17 Sustainable Development Goals (SDGs) (UN 2015). Since then, the role that Sustainability Education or Education for Sustainable Development (ESD) can play in achieving these goals, particularly within Higher Education Institutions (HEIs), has been increasingly highlighted (UNESCO 2017). This mandate for HEIs goes beyond preparing students for dealing with future occupational demands related to sustainability issues. It also implies addressing efforts to educate them as concerned, critical, and reflective citizens in relation to sustainability (Leal Filho et al. 2021). HEIs' networks at both global and regional levels are promoting change in HEIs' organizational culture to further advance the integration of the Agenda 2030 into their programs, activities, and discourses (Ruiz-Mallén

I. Ruiz-Mallén (✉) · M. Heras · R. Ribera-Fumaz · A. Corres
Internet Interdisciplinary Institute (IN3), Universitat Oberta de Catalunya (UOC), Barcelona, Spain
e-mail: iruiz_mallen@uoc.edu

H. March
Internet Interdisciplinary Institute (IN3), Universitat Oberta de Catalunya (UOC), Barcelona, Spain

Estudis d'Economia i Empresa, Universitat Oberta de Catalunya (UOC), Barcelona, Spain

P. Vare et al. (eds.), *Competences in Education for Sustainable Development*, Sustainable Development Goals Series, https://doi.org/10.1007/978-3-030-91055-6_10

and Heras 2020). These changes have mainly focused on incorporating sustainability content in the curricula, integrating sustainability as an outreach principle, examining sustainability-related issues from different research areas and designing and implementing campus operations based on environmental sustainability indicators.

Despite the progress made, significant challenges remain for advancing the incorporation of SDGs into university teaching and learning, such as improving teachers' competences in sustainability (Adams et al. 2018; Leal Filho et al. 2021). At the same time, there is a growing academic interest in exploring frameworks on sustainability competences for educators at higher education level. A recent systematic review has identified 11 studies that implement a variety of ESD competence models or frameworks targeted at HEI teachers (Corres et al. 2020). For example, the framework called RSP (*A Rounder Sense of Purpose*, Vare et al. 2019) reformulates the UNECE (2012) clustering of competences that reflect different learning experiences into the steps of a process that educators might follow (i.e., Integration, Involvement, Practice, and Reflection). Alternatively, the KOM-BiNE model (*Competences for ESD in Teacher Education*, Rauch and Steiner 2013) defines educators' competences according to their fields of action (i.e., Classroom instruction, Institutional participation, Reach society).

Further efforts have examined teachers' perceptions of ESD competences at higher education level. Leal Filho et al. (2021) conducted an international survey with HEI educators involved in teaching sustainability to find out to what extent they value ESD competences. They used a list of competences mainly based on UNECE (2012) and also relied on those suggested by Vare et al. (2019). Findings show how the teachers surveyed tend to value similarly both ESD knowledge-based competences and those more related to their teaching practice. For instance, they recognize the value of applying concepts to real-world problems as well as creating participatory and learner-centered learning opportunities. These findings suggest that they are aware of and have acquired ESD competences through their practice and own initiative (*op cit.*). But what happens with HEI teachers' perceptions of ESD competences in the case of teachers without or with little experience in sustainability education, particularly when there is already a well-established competence framework at the institution that is not specifically based on ESD?

This chapter aims to shed light on this question by examining the case of the Universitat Oberta de Catalunya (UOC). This online university based in Barcelona relies on competence-based learning. Although increasingly highly committed to sustainability issues, it developed its competence framework without fully embracing the ESD approach and, given the nature of its educational offer, the majority of the academic staff had little expertise or training in sustainability or ESD. In early 2020, the UOC launched an online training course for lecturers and professors to strengthen the integration of the Agenda 2030 and related competences into their teaching, with a particular focus on SDG 5: Gender Equality and UOC's *Ethical and Global Commitment* competence. The following discussion is based on a series of interviews with participant teachers and highlights some of the challenges and opportunities for engaging with the Agenda 2030 and ESD competences.

Building Bridges Between the UOC's Competence Framework and ESD

The UOC was created in 1995 and, as of 2020, has more than 70,000 students worldwide but mostly in Spain and Latin America. The university has seven Faculties: Business and Economics; Psychology and Education; Communication and Media; Arts and Humanities; Health Studies; Law and Politics; Informatics, Multimedia and Telecommunications Engineering. UOC's learning model is based on three main characteristics. Firstly, as an open university aiming to provide higher education to those who cannot attend classes in a specific location or specific schedule, the learning model uses asynchronous methodologies. Secondly, all teaching happens in a multi-lingual digital campus. Finally, as a university located within the European Higher Education Area, its learning approach is competence-based. Specifically, the university has a three-tier competence system:

1. specific competences related to a course subject,
2. transversal competences associated with doing a degree,
3. twelve transversal competences common to all degrees that are conceived as the UOC's competence framework: *Ethical and Global Commitment; Information Management; Planning and Organization; Team-Work; Oral and Written Expression; Foreign Language; ICT Use and Implementation; Creativity; Entrepreneurship; Learning to Learn; Resolution of Complex Situations; Analysis, Synthesis and Critical Thinking.*

Although the UOC's competence framework was developed without explicitly embracing the ESD approach, there is some correspondence with ESD competence frameworks for educators (Table 10.1). Four UOC transversal competences are understood in similar ways to four ESD competences defined by the RSP framework (ibid). Moreover, these four UOC competences hold direct and indirect linkages with the 12 ESD competences identified by the RSP framework.

UOC's competence of *Ethical and global commitment* (hereafter ethical competence), which has its equivalent in the *Responsibility* ESD competence (ibid), was the starting point for the online training course offered to UOC academic staff. The course objective was twofold: strengthening teachers' ethical competence and including Agenda 2030 in their teaching, in particular SDG5: Gender Equality. By following this rationale, the course was structured into three main modules named: (1) What can I do for Agenda 2030?; (2) Gender-sensitive subjects, and (3) *Ethical and global commitment* transversal competence. Each module included a couple of exercises on the related topic that the participants had to develop in the context of their teaching practice. Specialized readings and videos, some of them created for the course, were provided to support teachers in this endeavor. Also, participants were encouraged to participate in the online forum to share doubts and reflections with their peers.

Fifty UOC lecturers from across the different Faculties voluntarily participated in the 25-hour

Table 10.1 Correspondence between UOC competences defined by the UOC competence framework and ESD competences defined by the RSP framework (aroundersenseofpurpose.eu)

UOC competence and definition	ESD competence and formulation (from the RSP framework)	Related ESD competences (RSP)
Ethical and global commitment: Acting honestly, ethically, sustainably, and socially responsible in both academic and professional practice	*Responsibility:* The educator helps learners to reflect on their own actions, act transparently and to accept personal responsibility for their work	*Empathy*
Creativity: Applying creative thinking to provide improvements, solutions, relevance or differentiation in areas and situations of diverse complexity	*Creativity:* The educator encourages creative thinking and flexibility within their learners	*Attentiveness* *Empathy* *Futures* *Transdisciplinary*
Solving complex situations: Ability to solve complex situations by analyzing components, finding alternative ways to solve them, reaching consensus on their implementation, and evaluating the results of their implementation	*Action:* The educator helps the learners to take action in a proactive and considered manner	*Systems decisiveness* *attentiveness* *Empathy* *Engagement* *Futures* *Participation*
Analysis, synthesis, and critical thinking: Analyzing, synthesizing, and making judgments based on self-corrective and context-sensitive criteria	*Criticality:* The educator helps learners to evaluate critically the relevance and reliability of assertions, sources, models, and theories	*Decisiveness* *transdisciplinary* *systems* *Attentiveness*

training course in early 2020. We analyzed teachers' interaction along the course through the online forum and conducted semi-structured interviews with five of them after the course. All participants did the proposed exercises and actively participated in the online forum. For instance, during the first module participants reflected on how to embed the Agenda 2030 in their teaching practice and posted 177 messages. Only in a few cases (23 posts) did participant position themselves critically in relation to the program content. The main arguments were related to ontological inconsistencies between some SDGs, such as SDG 13 on Climate Action and SDG 8 on Decent Work and Economic Growth, as well as to tensions in the practical implementation of the Agenda 2030 and the ethical competence. The respondents also identified these aspects as difficulties for integrating the SDGs and ESD competences into the UOC's approach. The interviewees, in turn, detected opportunities; these are discussed below.

Tensions and Opportunities

The responding teachers identified conceptual and structural tensions that are particularly linked to higher education systems for practically addressing the SDGs through ESD competences.

Firstly, perceived limitations related to the adoption of plural and critical views required avoiding the normativity and fragmentation that are often adopted by HEIs when approaching the SDGs:

> The gender dimension was focused on gender parity, so it lacked an intersectional approach. What about the other social, cultural, environmental dimensions that go through the [gender] situation? Why such a partial approach? (Teacher #1).

Secondly, interviewed teachers perceived that the challenge of embedding Agenda 2030 into their teaching is more related to explaining content than to fostering specific competences:

> I do find more interesting to integrate the goals [SDGs] through the contents, which are of course linked to the competences. But I think that competences are the basis and the challenge relies on how you can move forward on competences to adapt the content and teaching to comply with the goals [SDGs] (Teacher #2)

Such a focus on content has been documented among HEIs worldwide, whatever the competence framework may be (Leal Filho et al. 2019).

Third, when trying to approach the SGDs from a competence-based learning perspective, interviewed teachers were challenged to deal with other individual dispositions beyond knowledge, such as values. Such challenge was particularly highlighted when trying to implement and develop the ethical competence:

> Does it exist a global, agreed ethics that we all share and can embrace to deal with the SDGs? Because there are contradictions between the ethical competence and some SDGs, such as in the case of the goal about growth [SDG8] (Teacher #3).

Indeed, the redefinition from content- to competences-based teaching approaches is still not solved in higher education. As Guerrero-Roldán and Noguera (2018) point out, teachers' misunderstandings about the difference between goals (i.e., what teachers expect to teach in terms of content and purpose) and resulting learning outcomes (i.e., the level of competence acquired) might be misleading when applied to competence-based learning, especially in the case of online environments. This confusion can make it difficult for teachers to find ways for addressing SDGs through specific ESD competences in their practice. But such a perceived limitation in addressing the Agenda 2030 within HEIs' teaching and learning from a competence-based perspective can lead us even to question the appropriateness of the approach. Why do UOC teachers who are used to work within competence-based learning schemes find it so challenging? Would such a barrier be related to the inherent normativity of the SDGs that is making teachers prioritize instruction over facilitation, teaching objectives over students' learning outcomes, and thus coming back to a teacher-centered paradigm? (Chaps. 2 and 6 specifically highlight potential flaws in the competence approach which may be relevant here). Or maybe is the focus on a content-based approach to address the Agenda 2030 more related to teachers' lack of long-term experience in sustainability education since the university is only recently engaging in these topics? In this regard, respondents questioned the validity of applying competences defined at the top in their subjects and called for more training courses on how to teach and assess learning about the Agenda 2030 from a competence-based learning perspective:

> More training would be needed on competences so we can be more explicit in the way we apply the goals [SDGs]. For instance in the case of empathy. As educators, first, we need to reflect on what this and other competences mean for us and in our context (…) Then we have to ask ourselves whether we are developing the competences that are assigned to us and whether these competences have any sense in our field or not (Teacher #4).

This demonstrates how simply describing a competence is insufficient in ensuring its acquisition as the labels and descriptors applied to competences often mask a great deal of complexity. Here the respondent highlights empathy, which has various interpretations and can be considered at multiple levels (see Chap. 2).

> The main challenge for me is the lack of teacher training to be able to teach and evaluate certain competences that teachers do not master because frameworks are often imposed from the top. (Teacher #3).

Despite these challenges, the teachers highlighted the strong institutional commitment made by UOC to promote sustainability education through the organization of the training course:

> Trying to integrate the goals (SDGs) at the institutional level through the competence of Ethical and global commitment, which is an important competence, is a brave step by UOC (Teacher #5).

They referred to some opportunities offered by the course to address the Agenda 2030 and integrate ESD competences in their teaching practice. The training allowed those teachers who were less familiar with the Agenda 2030 to reflect on how their subject areas and pedagogical strategies are already related to the SDGs:

> I had never thought that my academic and teaching work had anything to do with the Agenda goals, and the course allowed me to realize that it did. (Teacher #1).

Those teachers with more expertise in dealing with sustainability issues pointed out how their current teaching about some SDGs could be an entry point to be able to work them through ESD competences. For instance, and beyond the ethical competence, they mentioned existing linkages between their teaching practice and ESD competences that are currently implemented in their different subject areas or planned to be addressed in further curriculum development such as critical thinking, transdisciplinarity or action (Vare et al. 2019):

> I'm already working on the SDGs of zero hunger, sustainable cities and communities, gender equality and quality education because we teach an online Masters on food production and consumption (…) Transdisciplinarity is an objective for us; it is not a reality because we are currently working within an interdisciplinary approach (…) We are implementing competences related to taking action because practical reflection is also one of our goals. (Teacher #2).

Indeed, UOC teachers' previous experience on teaching about sustainability issues and implementing related competences is an opportunity for supporting the full development of competences on sustainability within the institution. Reflecting on the potential of experienced teachers in ESD at HEIs, Leal Filho et al. (2021) suggest emphasizing their role as trainers and mentors through engaging them in the design and implementation of training methods for building capacities and motivating those teachers with less expertise or interest in sustainability issues.

Final Remarks

The online training course represents an essential institutional effort to guide UOC academic staff in embedding Agenda 2030 in their teaching while developing related sustainability competences in themselves and their students. Despite the fact that the UOC's competence framework was not inspired by ESD approaches, some ESD competences partially fit within the UOC pre-established framework, such as the ethical competence addressed during the online training course.

Having a pre-established competence framework does not necessarily mean resistance to the implementation of ESD competences at HEIs. Conceptual and structural tensions related to the application of competence-based learning to teach about SDGs seem to be more challenging for UOC teachers in this regard. In addition, many of them lack long-term experience in the field of sustainability education. The language of competences can help these teachers in addressing the Agenda 2030 SDGs because it is the language already used for curriculum and pedagogical strategy development at UOC. The competence-based approach can also help them teach about sustainability issues in a more systemic and reflexive way but more training on sustainability competences seems to be required. Further efforts on developing these training courses should provide clear guidance on how to address sustainability using a competence-based approach that goes well beyond content-based teaching.

Acknowledgements We are thankful to the UOC team who organized the online training course and to those lecturers and professors who participated in the interviews. This research was funded by European's Union Erasmus+ program under grant agreement No. 2018-1-UK01-KA203-048121 "RSP II: A Rounder Sense of Purpose II (RSPII)" and the Urban Transformation and Global Change Laboratory (TURBA Lab). I. Ruiz-Mallén gratefully acknowledges the support of the grant RYC-2015-17676 funded by MCIN/AEI/ 10.13039/501100011033 and by "ESF Investing in your future". A. Corres has been funded by the Universitat Oberta de Catalunya through a Doctoral Thesis Grant.

References

Adams, R., Stephen, M., & Boom, K. (2018). University culture and sustainability: Designing and implementing an enabling framework. *Journal of Cleaner Production, 171*, 434–445. https://doi.org/10.1016/j.jclepro.2017.10.032

Corres, A., Rieckmann, M., Espasa, A., & Ruiz-Mallén, I. (2020). Educator Competences in Sustainability Education: A Systematic Review of Frameworks. *Sustainability, 12*(23), 9858. ; https://doi.org/10.3390/su12239858

Guerrero-Roldán, A., & Noguera, I. (2018). A model for aligning assessment with competences and learning activities in online courses. *The Internet and Higher Education, 38*, 36-46. https://doi.org/10.1016/j.iheduc.2018.04.005

Leal Filho, W., Shiel, C., Paço, A., Mifsud, M., Veiga Ávila, L., Londero Brandli, L., Molthan-Hill, P., Pace, P., Azeiteiro, U.M., Ruiz Vargas, V., & Caeiro, S. (2019). Sustainable Development Goals and sustainability teaching at universities: Falling behind or getting ahead of the pack? *Journal of Cleaner Production, 232*, 285–294. https://doi.org/10.1016/j.jclepro.2019.05.309

Leal Filho, W., Levesque, V.R., Lange Salvia, A., Paço, A., Fritzen, B., Frankenberger, F., Damke, L.I., Brandli, L.L., Veiga Ávila, L., Mifsud, M., Will, M., Pace, P., Azeiteiro, U.M., & Orlovic Lovren, V. (2021). University teaching staff and sustainable development: an assessment of competences. *Sustainability Science, 16,* 101–116. https://doi.org/10.1007/s11625-020-00868-w

Rauch, F., & Steiner, R. (2013). Competences for Education for Sustainable Development in Teacher Education. *Center for Educational Policy Studies Journal, 3*(1), 9-24. https://doi.org/10.25656/01:7663

Ruiz-Mallén, I., & Heras, M. (2020). What Sustainability? Higher Education Institutions' Pathways to Reach the Agenda 2030 Goals. *Sustainability, 12*, 1290. https://doi.org/10.3390/su12041290

UN. (2015). *Resolution adopted by the General Assembly on 25 September 2015. 70/1. Transforming our world: the 2030 Agenda for Sustainable Development.* Retrieved November 5, 2020, from https://www.un.org/ga/search/view_doc.asp?symbol=A/RES/70/1&Lang=E

UNECE. (2012). *Learning for the future. Competences in education for sustainable development.* United Nations Economic Commission for Europe. Retrieved November 4, 2020, from https://www.unece.org/fileadmin/DAM/env/esd/ESD_Publications/Competences_Publication.pdf

UNESCO. (2017). *Education for Sustainable Development Goals: Learning Objectives.* UNESCO.

Vare, P., Arro, G., De Hamer, A., Del Gobbo, G., de Vries, G., Farioli, F., Kadji-Beltran, C., Kangur, M., Mayer, M., Millican, R., & Nijdam, C. (2019). Devising a competence-based training program for educators of sustainable development: lessons learned. *Sustainability, 11*(7), 1890. https://doi.org/10.3390/su11071890

Breaking the Mold: Educators as Agents of Change

11

Francesca Farioli and Michela Mayer

Abstract

Educators, including those who are involved in education for sustainability (EfS), are not always aware of their role as agents of change. Yet education is not only a fundamental drive for creating and/or transmitting the values that shape a society, but also for its transformation, since it helps to build worldviews, values, competences, and actions that can lead society in the direction of a sustainable future. Educational literature focuses mainly on the role of students as future citizens and agents of change and on the competences they must acquire. To achieve this role, educators need to be aware that they are change agents themselves and therefore need to acquire relevant competences. This chapter reflects on the experience of four competence-based training courses carried out in Italy, involving both formal and non-formal educators through both face-to-face and distance modalities, and explores the way in which the use of an integrated framework helped to enhance the educators' awareness of their own role as agents of change.

Keywords

Change agents · Non-formal educators · Action research · ESD competences · RSP

F. Farioli (✉) · M. Mayer
IASS, Italian Association for Sustainability Science, Rome, Italy

...he views the world as a 'system of systems' where each system conditions the others and is conditioned by them. *(Calvino 1988, pp. 105–106)*

Introduction

"Educators are powerful change agents who can deliver the educational response needed to achieve the SDGs. Their knowledge and competencies are essential for restructuring educational processes and educational institutions towards sustainability "(UNESCO 2017 p. 51). This restructuring process implies a whole paradigm change which encourages a shift from a transmissive expert-based teaching to a transformative community-based, learning.

The above concern inspired the Rounder Sense of Purpose (RSP) project, which sets out to develop and test a framework of competences that an educator capable of promoting such a transformation should have (Vare et al. 2019; see also Chap. 5).

Although education for sustainability (EfS) widely recognizes that educators themselves are 'agents of change', the challenge remains on how to break the educational system's "mold" and convince the educators of their role: to be the innovators themselves and not only the transmitters of innovation. Educators could fully perform this role.

As underlined by Sterling (2011), the change that is required is not just doing what we do better, i.e. substituting one element with another, or improving one or more aspects, but it is changing the way we think about our world and society, a

P. Vare et al. (eds.), *Competences in Education for Sustainable Development*, Sustainable Development Goals Series, https://doi.org/10.1007/978-3-030-91055-6_11

'second order change' which involves a "*Deep structural shift in the basic premises of thought, feelings and actions, a shift of consciousness that dramatically and permanently alters our way of being in the world*" (O'Sullivan and Morrell 2002, p.18). For that to happen Education itself needs to change and educators have to become agents of change themselves and contribute to social transformation.

This chapter reports on the experiences of four different competence-based training courses, carried out in Italy, involving both formal and non-formal educators, through both face-to-face and distance modalities, and explores the way in which the use of the RSP integrated framework helped to enhance the educators' awareness of their role as agents of change.

Who Are the Educators Promoting Change?

By the term educators we mean both teachers and educators of the non-formal sector. When one talks about change agents, it mainly refers to learners as future citizens and changes are mainly those related to behavior. However, in order to promote transformative changes among their learners, educators need to be and act as agents of change themselves. But on what occasions do educators become conscious agents of change?

Where education literature talks about agents for change, it tends to focus on innovation, generally in the technological and/or digital sense: a rather limiting definition. In present-day schooling, teaching is conceived more as an 'adaptation' to the current society, even if, on the contrary, society is in continuous evolution and the agent of change is therefore only a carrier of technological or disciplinary innovation. Even in this limited conception of change agent, it is recognized that teachers need other competences such as communication skills, empathy, and/or leadership.

In our vision educators, instead, are agents of change when they propose themselves as a facilitator, providing food for thought on society and on the planet and by unpacking and negotiating that very same knowledge that is now considered indisputable (e.g. what is considered Science or progress, as proposed by post-Normal science, see Chap. 8).

Even though they are not always aware, teachers, or non-formal educators, are both agents that promote a change in their learners as well as subjecting themselves to change: in this 'structural encounter' (Maturana & Varela 1987) and by playing the role of a facilitator, educators inevitably change and alter themselves, "in a process of transformation in coexistence during which previous participants' conditions and context are altered as a result of their interaction," (Souza et al. 2019, p. 1609) and as expressed by Calvino in our opening quotation.

The courses reported in this chapter were aimed at triggering, among in-service teachers and educators engaged in EfS actions, an individual and then collective reflection on their role, on the change they would like to induce in their learners and on what they themselves are experiencing, by not only being aware of living in a 'system of systems' but by being able to impact on it.

The Italian Context for Education for Sustainability

Over the last 20 years in Italy, the changes in the education system toward EfS have been conflicting and not effective. Starting from the 1990s up to the first decade of 2000, and mainly thanks to the input of the Ministry of the Environment, a national plan (Italian National System on Environmental Information, Training and Education—INFEA) was launched to support the creation of a 'system' that could connect and foster the implementation of the many experiences existing all over the country. Since that positive experience, unfortunately, many 'threads of the network' have been severed, threads that have somehow flowed back into non-communicative fragments; nonetheless professionalism and a widespread interest in the evolution of EfS remain. The INFEA System had in fact allowed a common evaluation on the 'quality' of education for sustainability and also a methodological debate on the necessary competences of EfS practitioners, a confrontation that has been lost over time.

This does not mean that, currently, in Italy there are no peaks of excellence—such as highly innovative schools and networks of schools—or that there is no strong interest in EfS. This is especially strong in compulsory education, finding confirmation in an abundance of proposals for educational activities, offered both by local institutions (regions and municipalities), Park Authorities and Environmental Agencies, and finally by national and local associations.

In general, however, the Italian context (cultural, political, and institutional) does not recognize teachers and educators as change agents and they are often asked to play an 'adaptive role' for the maintenance of the knowledge transmission chain for what society considers to be valid and useful to the preservation of a social, economic and cultural *status quo* avoiding, as much as possible, a real change in how we think and act.

In Italy the mastery of disciplinary knowledge has always been considered more important than pedagogical and methodological competence: for lower secondary school teachers (11–14 years of age), and upper secondary school teachers (15–19 years of age), no specific degree is expected, except for the disciplinary Master's degree.

This situation is also partly shared by non-formal educators—whose methodological and pedagogical professionalism, built through individual or associative paths, is often unrecognized. Educators present schools and citizens with educational activities, both in the environmental field (for example, outdoor education) and in the field of sustainability (e.g. lessons or activities on waste or renewable energy), as well as authentic educational projects to be integrated with school work. However, their actions are often strongly conditioned by the appointing body (the schools themselves, but also the public administrations or companies that manage the services such as water, waste, energy) that often fails to ask them to carry out transformative educational processes but rather to focus on random interventions on specific content.

Insufficient investment in education, an increased bureaucracy over the last 20 years, and above all a weakened social image—i.e. school no longer guarantees more or better job opportunities—have made the work of teachers more difficult and teachers themselves have been less and less interested in the role of change agents.

The main point of contact between the challenges that teachers and non-formal educators face in their commitment to a 'sustainable education' is that of having to fight, as Sterling (2010) suggests, a highly resilient organizational and cultural context; one that is resistant to change, which does not require educators as 'conscious builders of the future' nor 'cultural mediators' between knowledge and society, but as disciplinary experts, capable of transferring specific contents in any situation. And this, in spite of the fact that the European document on key competences (Council of Europe 2018) has been disseminated in all schools, and that the national strategy toward the 2030 Agenda considers as an 'indicator', precisely the 'transition from knowledge to competences' (Italian Strategy for the 2030 Agenda, 2017).

Hence the 'professional discomfort' experienced by both teachers and educators, finding in the RSP competences a useful tool for reflection and research that can support them while abandoning the role of 'experts' in which they have been confined.

In this rather disheartening institutional framework, an important step in 2019 was the recognition of the *'educational mission for the environment and sustainability' of* SNPA[1] (consisting of ISPRA—National Institute for Environmental Protection and Research and the Regional Agencies for Environmental Protection—ARPA). ISPRA and some ARPA staff members were also key players to the INFEA system. A possible strategic element of transformation, to which a proposal to spread the RSP competences in Italy immediately connected.

[1] SNPA: National System for Environmental Protection.

Piloting the RSP Competence Model

The existence of a shared context and problems at a national level allowed us to offer fairly similar paths for the four training courses we devised—two in 2017/2018 and two in 2020—to validate and disseminate the RSP model and to identify a common goal: building, among educators and teachers, a greater awareness of their role as agents of change, thus proposing an action and reflection path on the necessary competences. In Table 11.1 we have summarized the main objectives and characteristics of the four courses.

In the first two courses, carried out between 2017 and 2018, the main objective was to test the validity of the RSP model, discussing it with experienced teachers and educators. The course had the additional aims of developing, with participants' help, data collection tools as well as reflecting on the professionalism that education for sustainability requires as well as on their acting, sometimes not consciously, as agents of change.

The first course was designed together with the University of Florence (working with Professor Giovanna Del Gobbo) and carried out at the Villa Demidoff Environmental Education Center (a center that carries out EfS activities for schools and teachers in the metropolitan city of Florence). The course, formally recognized by the school authority as in-service training, was aimed at teachers of different levels and types of school, from primary to upper secondary: almost all were experienced teachers, interested in EfS, often with roles of coordination of the EfS activities in their school. The course did not include a formal final assessment, only a certificate of attendance.

For the second course we involved an environmentalist association, Legambiente, that has a high profile nationally, in particular the division that deals with education and training with participating educators in different parts of Italy. Again, in their case no final assessment was required.

The third course, held between March and September 2020, was very different: the goal was no longer to validate the RSP model but to test its effectiveness on a national level. The National System for Environmental Protection (SNPA) had seen the RSP model presented publicly at the end of Phase One of the project in 2018; they saw in this an opportunity for a national training and professional development course on EfS based on the RSP competences framework with funding requested from the Ministry of the Environment. The course was open to regional officials dealing with the organization of EfS activities as well as communication on sustainability, to active educators in the environmental education centers, and to technicians from Regional Agencies of Environmental Protection who operate in schools when needed. The course,

Table 11.1 Key characteristics of the four courses

Main aims	Participants	Hours	Partner Institutions
Validation of the RSP palette	15 teachers from various school levels	40 h including remote work	Florence University and Villa Demidoff EE Center
Validation of the RSP palette and the evaluation of levels	15 experienced environment and sustainability educators	40 h including remote work	Legambiente national association
Testing the effectiveness of the RSP proposal at national level	84 educators, technicians, supervisors	6 modules, remote work, and 4 lab hours, dedicated to RSP competences and action research	SNPA
Design a collaborative action research on RSP educator competences	20 educators, video makers	8 h interactive face-to-face training, 2 remote focus groups	ARPA Friuli Venezia Giulia—Environmental education Laboratory of Friuli Venezia Giulia Region (LaREA)

available for a maximum of 100 people with participants from each Italian region and designed to be carried out remotely, also included face-to-face meetings. The COVID emergency transformed face-to-face meetings into four remote labs for the entire day, each aimed at 20 people coming from all over the country and with different job positions.

The fourth course, held between November 2019 and December 2020, was aimed at exploring how a collaborative action research on RSP educator competences could be designed and carried out with a heterogeneous group of participants and whether this experience could enhance the quality and the effectiveness of educational proposals made by them, as well as improving the collaboration within the group, in this case the local educational network promoted by LaREA (see Table 11.1). Participants on this course were educators from local cooperatives and video makers who work with schools to promote 'image culture' in the media context. The main challenge was how to design a collaborative action research with people who share the same values, visions, overarching aims in Education while using different tools and 'languages' in their practices.

Despite the differences in specific objectives including specific requests from teachers/educators themselves, some elements remained the same in all four courses:

- Firstly, the choice to present EfS as a Transformative Education (Wals 2007) within the overarching framework proposed by the 2030 Agenda, and to consider RSP competences as a constant thread, that can link visions of the world and concrete actions for sustainability, in a training path that focuses on self-reflexivity
- The starting point (including on remote learning courses) is always making contact with oneself, with one's own values, experiences, professionalism, as people and as possible agents of change. For this purpose, we used an 'environmental autobiography,' asking participants to get in touch with one's inner self through metaphor and image. When it was possible to share the autobiographies (both on site and partially via the remote lab), the comparison allowed us to also share our personal "mission" and therefore different visions of one's own professionalism
- The competence model was presented as a whole, in order to show its potential to offer '*a rounder sense of purpose*' to the many practical activities often considered as EfS. We never lost track of the RSP palette as a whole (see Chap. 5), although we examined in depth certain competences in the on-site courses (and all of them in the online courses), offering ideas for reflection but also asking for individual and group working on concrete activities. We experimented with and reflected on each competence's learning objectives and tried to use the underpinning components as clues, traces that would allow us to follow the path of the development of competences
- We asked participants to create an individual portfolio in which to collect, in addition to one's own autobiography, starting point to follow the transformation, the "stories" to write and present: significant experiences of the competences "acted on" in their own classes or in their own working groups (Farioli & Mayer 2020)
- We dug into what was already there (many of our 'learners' were experienced educators) and invited them to look within themselves and compare their own practices and thus their own theories. *Looking at oneself and reflecting and engaging with others* was the focus of the work for gaining awareness that we proposed in all four courses. Different tools and activities were used: group discussion; portfolio; self-evaluation; peer evaluation—but always with the same goal: to explore one's own vision of the role of change agent and compare it with those of others
- The first three groups were asked for a final self-assessment on the 12 competences, following the three levels of competence proposed by the RSP project (Vare et al. 2019). In the final discussions, held at a distance both

spatially and temporally, and in the final questionnaire we asked, we collected *evaluations and judgments* on how participants perceived the course but also reflections on *what had been learned.*

What Have We Learned?

The first two courses in 2017–2018 were set up as a collective action research where experts, educators, and teachers gathered data and reflected together on the RSP model. On the path of researching together, the activities we carried out were, at the same time, shared examples of data collection (on educational methods, values, management of actions and emotions, capabilities, etc.) and opportunities to think about one's role as agents of change. We asked participants to use competences as a 'magnifying glass' to examine one's own professionalism (self-evaluation) and to compare it with that of others (peer-evaluation). From the work we did together, we collected material to reflect on as well as to suggest future action. Two years later, in 2020, we asked participants what remained of that joint work experience, what changes it had triggered; the following are some of the answers we gathered:

> It allowed me to 'systematize' the concepts I had acquired in my experience as a teacher, and also give importance to non-disciplinary, 'transversal' aspects, ... enhancing this type of approach... (AD, Teacher)

> During the planning and development of my educational activities, I tried to pay more attention to some competences that I had underestimated before the course (MS, Educator)

> In these two years I have turned my attention to the sustainability of my teaching, revisiting it in light of competences such as: ' empathy', 'values', 'transdisciplinarity' (MT, Teacher)

> Obviously the RSP model should not be applied rigidly, but it should work as a background within which you can move, even with freedom and creativity, so as to avoid it becoming a cage (TD, Educator)

> All the competences proposed by the RSP project challenge me daily. I must say that being able to actually put them into practice is an ongoing process, (and not always a successful one) (ES, Educator).

Even in the remote course we tried to maintain a researching approach, but the involvement of the learners was unavoidably reduced: we collected the satisfaction data at the end of the course, we tried to follow the learning path through the 'assigned tasks', we also tried to understand what the learners' 'starting competences' were, both through CVs and autobiographies, and by building a Likert scale that would highlight the 'educational beliefs', i.e. attitudes toward more transmissive or more transformative educational approaches. Within the four remote labs carried out at the end of the course, we conducted focus groups of about ten people where we discussed the RSP competences and their possible use as 'drivers of change', using, in small groups, assigned tasks as food for thought. After making these tasks anonymous, we used them as case studies relating to the use of the RSP competences, discussing the possibility of recognizing the competences when they are acted on and the need to use them if one is willing to be an agent of change.

To conclude, the RSP competences were an important research and reflection tool not only for IASS but for much of the environmental education and sustainability world in Italy: the collaboration established in these 2 years with SNPA, with INDIRE—National Institute for Documentation, Innovation and Educational Research— and with other national and regional bodies, testifies to the interest that the proposal has aroused and to the possibilities of future impact both on the national school system and on the regional networks as well as on the environmental education centers. Our experience can be summed up in the words of Paulo Freire (2007):

> No one educates another, no one educates himself alone: people educate themselves together throughout the world.

References

Calvino, I. (1988). *Six Memos for the Next Millennium*, Cambridge, Harvard University Press

Council of Europe Recommendation of 22 May 2018 on key competences for lifelong learning (2018/C 189/01)

Farioli F. & Mayer M. (2020). *Transformative Evaluation for Sustainability Educators: The Experience of a Competence-Based Educational Research Project.* 2nd GUNi International Conference on SDGs: Higher Education & Science Take Action. Summary Report. pag.101- 104 ISBN978-84-09-23375-5 Downloadable open-content version available at: www.guninetwork.org

Freire, P. (2007). *Pedagogy of the Oppressed.* New York: Continuum

Italian Strategy for the 2030 Agenda (2017), https://www.minambiente.it/sites/default/files/archivio/allegati/sviluppo_sostenibile/Italy_in_a_glance_VNR.pdf, retrieved on 07/02/2021

Maturana, H. R., & Varela, J. F. (1987). *The Tree of Knowledge: The Biological Roots of Human Understanding.* New Science Library/Shambhala Publications

O'Sullivan, E., A. Morrell(Eds) (2002). *Expanding the boundaries of transformative learning: Essays on theory and praxis.* New York: Palgrave

Souza, D.T., Wals A. E. J. & Jacobi P.R. (2019). Learning-based transformations towards sustainability: a relational approach based on Humberto Maturana and Paulo Freire, *Environmental Education Research*, 25:11, 1605-1619, DOI: https://doi.org/10.1080/13504622.2019.1641183

Sterling, S. (2010). Learning for resilience, or the resilient learner? Towards a necessary reconciliation in a paradigm of sustainable education, *Environmental Education Research*, 16:5-6, 511-528, DOI: https://doi.org/10.1080/13504622.2010.505427

Sterling, S. (2011). Transformative learning and sustainability: Sketching the conceptual ground. *Learning and Teaching in Higher Education* 5: 17–33.

UNESCO (2017). *Education for Sustainable Development Goals. Learning Objectives.* Paris: UNESCO. http://unesdoc.unesco.org/images/0024/002474/247444e.pdf (Lead authors: Marco Rieckmann, Lisa Mindt, Senan Gardiner)

Vare, P., Arro, G., de Hamer, A., Del Gobbo, G., de Vries, G.; Farioli, F., Kadji-Beltran, C., Kangur, M., Mayer, M., Millican, R., Nijdam, C., Réti, M., Zachariou, A. (2019). Devising a Competence-Based Training Program for Educators of Sustainable Development: Lessons Learned. *Sustainability*, 11, 1890

Wals, A. E. J. (2007). *Social Learning towards a Sustainable World* (A. E. J. Wals (ed.)). Wageningen Academic Publishers

12 Framing the Frames: Integrating an ESD Approach Into an Existing National Framework

Mónika Réti, Edit Lippai, and Márk Nemes

Abstract

Mainstreaming ESD competences can occur in various forms; one approach is to integrate them into an existing structural, theoretical and operational framework. This chapter reviews the multistakeholder process involved in integrating an ESD competence framework within an existing general framework used for assessing in-service teachers' competences in Hungary. After introducing the process, the chapter explores the structural challenges. These are partly linked to adapting a general pedagogical model for a national system, which has been designed with specific purposes and criteria while still being able to arrive at a concise, measurable, certifiable and documentable framework. Another important set of challenges is connected to accessibility. While addressing these challenges, the question of authenticity is strongly present: is it possible to deconstruct and rebuild a framework without losing its key values and ideals?

Keywords

Competences · Mainstreaming · Assessment · Authenticity · Sustainability

Introduction

Mainstreaming an ESD educator competence framework involves challenges regarding the national policy framework in sustainability and general and higher education, but also those of harmonising the framework approach with structural characteristics of in-service teacher assessment, the advancement system, evaluation and professional development. In Hungary, the policy framework for ESD is well elaborated and connate. Political will was also present to introduce ESD competences in the general assessment of educators working in general education in the country. The intention was to set up indicators in support of ESD competences that fit into the existing framework and that are adaptable for preparing and assessing pre-service and in-service teachers. The first stage of the process was to elaborate the indicators and then provide descriptions for assessing practising educators. The second stage was to apply the system to initial teacher training including compulsory curricula and output requirements.

M. Réti (✉) · E. Lippai · M. Nemes
Hungarian Research Teachers Association,
Budapest, Hungary

P. Vare et al. (eds.), *Competences in Education for Sustainable Development*, Sustainable Development Goals Series, https://doi.org/10.1007/978-3-030-91055-6_12

While elaborating the standards and indicators for ESD competences, they had to be kept both specific and identical with the original model, but they needed to be formulated so as to be agile enough to be meaningful for each educational professional involved in the assessment process: kindergarten teachers, primary school teachers, all subject teachers in secondary education, special needs teachers, speech therapists, psychologists, even principals and supervisors. This implied a set of challenges to establish an accessible, comprehensible, legible, yet precise and distinctive collection of provable indicators that could be applied at six different stages of educators' professional development according to the national career model.

General structural challenges involved adapting a general pedagogical model for a national system with specific purposes and criteria, resulting in a concise, measurable, certifiable and documentable framework. Moreover, this general pedagogical model of educator competences in Hungary that depicts the characteristics of an 'ideal educator' (Kotschy 2014), using indicator clusters and standards defined at six stages (Kotschy 2011), happens to overlap significantly with the model that served as a basis for introducing ESD competences. This resulted in challenges connected to integrating a holistic ESD competence model to a previous (also holistic) general model while avoiding repetition and redundancy.

Educator Competences: Weaving the Old with the New

The competence definition in the national qualification competence framework in Hungary is in harmony with international concepts (Demeter 2006). It applies a holistic approach, based on complex, long-term pedagogical research and analysis (Kocsis et al. 2012). However, in Hungary, educator competences go beyond a theoretical framework: educator appraisal has an impact on the educators' salaries and may influence their career opportunities. While schools can choose the criteria for teacher appraisal, evaluation of the selected criteria is strictly regulated (Réti 2019). Educators may advance in the career model if they undergo the qualification process against all indicators listed in the competence framework (Kotschy 2011). Teachers are regularly invited to upload an e-portfolio based on which of their professional activities are to be evaluated, thus their advancement is stated. Between the introduction of the model in 2013 and 2021, over 88,000 teachers have taken part in the qualification process that represents roughly 50% of the educators employed in general education. This underlines the level of responsibility involved when proposing any changes to the model as they will affect the entire educator community in the country.

After the political decision to introduce ESD competences to the national qualification competence framework in 2017, the Ministry of Human Capacities of Hungary (responsible for general education in the country) examined several competence models as well as widely recognised international approaches such as the Learning for the Future model (UNECE 2011). Some of these models had contributors from Hungary or pilots involving Hungarian teachers including action-oriented models such as KOM-BiNE (Rauch et al. 2008; Rauch and Steiner 2013) and dynamic models of ESD competences (Sleurs 2008). Finally, based on experts' suggestions, decision-makers opted for the Rounder Sense of Purpose[1] (RSP) model (Vare et al. 2019) as it was concise and seemed possible to be consistent with the approach of the qualification framework introduced in 2013. The national qualification model contains 62 indicators in eight competence areas and these indicators correspond well with the so-called underpinning components of the RSP competence model. This served as a basis for future work.

In 2018, a multistakeholder expert board was established to examine the possibilities to complement the national qualification framework with the RSP model and to elaborate supplementary standards and indicators based on which educators will be assessed in the coming years.

[1] https://aroundersenseofpurpose.eu.

This group contained independent researchers, university educators involved in initial teacher trainings, school leaders, teachers from kindergarten to secondary levels (across all subject domains defined in the national core curriculum in Hungary), in-service teacher trainers, teacher mentors and evaluators. Their work was organised by the Educational Authority of Hungary with altering in-person and online remote working sessions. The experts consulted the RSP model, then associated the underpinning components of the RSP competences to indicators in the national qualification competence framework. This resulted in a matrix where 14 out of 53 underpinning components remained unmatched. In case of some competences such as *Systems* and *Futures* nearly all possible links were missing, which created a clear call for including them directly as competence indicators. However, in other cases although matches were identified, the context of the specific indicator and the underpinning component was slightly (or more markedly) different. In the case of *Criticality*, for example, some underpinning components showed up to seven partial matches. Debates were raised whether partial matches are acceptable at all or whether a sum of partial matches can make up for a full match in the case of a competence. These decisions were considered individually in the case of all underpinning components where partial matches were involved. Finally, four indicators were added to the existing national framework: two of them includes *Systems* and *Futures* (which were completely missing from the original set of indicators) and two of them sum up elements from competences such as *Attentiveness* where experts could not accept partial matches as appropriate links. As a probe, the reverse exercise was also performed as seen in Table 12.1.

The process raised several critical questions. The major question persisted the process itself: is it possible at all to break down and restructure a concise competence framework in a way that the result is coherent with the original one? And if so: what processes and considerations may guarantee the congruence and the coherence of the model and the resulting complementary framework? Moreover, what can be done to guarantee that the values and ideals within and beyond the model are treasured in the resulting derivation of the model? Finally, to what extent are a framework's elements (i.e., underpinning components) transformed when inserted in a different competence framework? Will they preserve the same meaning or will the converted (or metamorphosed) meaning change the essence of the original model?

These questions remain open despite the fact that the two models match well, and in spite of the efforts made by the expert group to conserve the context of the RSP model while complementing the national qualification model with the ESD competences.

Pedagogical Questions

Earlier pilots with ESD educator competences highlighted the importance of accessibility, ensuring that teachers understand, work with and follow their development in the competence framework (Gugerli-Dolder and Rauch 2012). In Hungary, educators realise the necessity of competences—even if they admittedly do not possess all of them—and understand the concept of competences (Korpics 2015). As was expected, the introduction of ESD competences did not incur resistance from them. In the first wave, 15,062 teachers qualified successfully against the ESD competences. Their overall results were high, which suggests that their ESD competences are satisfactory. But how does this translate into classroom practice? And does the proven possession of the competence component mean that they have a good command of ESD? Or does this reflect that they take part in large-scale national ESD events and profit from those? And if the latter is the case, will this result in meaningful changes in students' ESD competences? More research will be needed to answer these questions.

Table 12.1 Matches between the national qualification competence framework and the RSP model (X indicates where a match was detected)

Competence area	Indicator	Matches with the RSP model
Professional tasks, academic, subject and curriculum-related knowledge	1.1. The teacher's pedagogical activities reflect a well-established academic knowledge	X
	1.2. Knows the policy documents issued by the Government and the Minister of Education applied in their institution regulating the content and operation of pedagogic work as well as the relevant content for their subject/expert area in their school's pedagogical programme	
	1.3. Knows and consciously applies the connections between their subject field and other cultural domains or pedagogical areas	X
	1.4. Knows and consciously applies learning processes, teaching methods and tools related to their subject field	X
	1.5. Knows important information sources relevant for their subject field, aware of their use, reliability and ethics in pedagogy	X
	1.6. Using appropriate terms related to pedagogical situations	X
Designing pedagogical processes and activities and self-reflections related to their realisation	2.1. While planning, considers local curricula, content regulations and inner expectations by their school, as well as development goals of individuals and groups taught by them	X
	2.2. Plans strategy, processes, work forms, methods and tools in a systemic approach	X
	2.3. Developing activities of their students is a priority in their pedagogical planning	X
	2.4. Builds on opportunities offered by social learning in their planning activities	X
	2.5. Plans a differentiated teaching-learning process, tailored to individual needs of optimal development of their students	X
	2.6. Includes motivation and developing motivation of students in their planning	X
	2.7. Includes extracurricular learning opportunities in the learning processes while planning	X
	2.8. Designs assessment methods and tools for individuals and groups	X
	2.9. Includes students in planning the teaching-learning process also considering their developmental stages	X

(continued)

Table 12.1 (continued)

Competence area	Indicator	Matches with the RSP model
Supporting learning	3.1. In supporting learning, builds on individual goals and needs of students as well as the group's characteristics	X
	3.2. Considers the actual physical and emotional state of students	X
	3.3. Raises and keeps students' curiosity	X
	3.4. Creates a safe and calm learning environment	X
	3.5. Identifies and appropriately treats learning difficulties during the learning process	X
	3.6. Motivates students for sensible, critical and ethical use of info-communication tools in learning	X
	3.7. Develops students' learning skills	X
	3.8. Provides guidance and appropriate tools for individual/autonomous learning	X
	3.9. Considers students' making mistakes as a part of the learning process and reacts to them in a way to improve understanding	X
	3.10. Supports autonomous thinking of students, acknowledging their initiatives and ideas and integrating those in the learning process	X
Developing students' personalities, individual approach, pedagogical knowledge (and readiness) to inclusively teach special needs students, students with unfavourable backgrounds, students with behavioural, socialisation, learning difficulties	4.1. Pays special attention to the learners' cognitive, social, emotional and physical needs in the learning process	X
	4.2. Consciously creates pedagogical situations supporting the complex personal development of the learners	X
	4.3. Respects the personalities of learners, consciously seeking their values, and takes an inclusive and responsible approach to the learners	X
	4.4. Introduces general human, European and national values identified for the specific age group in curricula and content regulating documents to learners	X
	4.5. Promotes conscious value choice and establishing own values to learners	X
	4.6. Consciously applies pedagogical–psychological methods to know their learners in a multi-dimensional way	X
	4.7. Recognises difficulties concerning personal development or learning of their students and is able either to efficiently help them or to seek other experts' assistance	X
	4.8. Recognises the promise of talent in the learners and consciously helps to develop that	
	4.9. Finds ways to individually develop learners in an inclusive environment	X

(continued)

Table 12.1 (continued)

Competence area	Indicator	Matches with the RSP model
Initiating forming groups and communities of students, providing equal chances, openness to social and cultural diversity, integration students, acting as form-teachers	5.1. Bases the development of the learner groups on the knowledge of group dynamics of community building and the individual and group needs and characteristics	X
	5.2. Creates the conditions for the development of collaborative skills during the designed educational situations	X
	5.3. Appropriately and efficiently applies methods of preventing and dealing with conflicts	X
	5.4. Promotes exchange of idea between students, develops their communication skills and the culture of discussion and argumentation	X
	5.5. Creates pedagogical situations to improve the social responsibility of learners	X
	5.6. Considers characteristics originating in different social and cultural backgrounds of students and transfers them as values towards learners	X
	5.7. Educates students to accept, respect, mutually support each other without prejudice or biases	X
	5.8. Their pedagogical activities reflect the knowledge of developmental psychology in the age group	X
Continuous assessment and analysis of pedagogical processes and the personal development of students	6.1. Applies curricula adapted to the learners individual pedagogical–psychological needs	X
	6.2. Applies various pedagogical assessment methods, purposefully applying forms of diagnostic, formative and summative assessment in the educational process	X
	6.3. Uses different assessment, monitoring and evaluation tools relevant for subject area and the specific pedagogical situation	X
	6.4. Chooses or prepares appropriate assessment or evaluation tools in line with pedagogical aims	X
	6.5. Provides individual assessment to learners	
	6.6. Analyses and regularly uses the results of students' assessment in establishing pedagogical developmental aims and tasks	X
	6.7. Considers the expected effect of applying specific assessment tools on the personal development of learners when using those	X
	6.8. Provides regular, appropriate and objective feedback to students	
	6.9. Improves the self-evaluation of learners	X
	6.10. Introduces the assessment and evaluation system applied in line with the school's pedagogical programme in the beginning of the teaching-learning process to the learners and parents	

(continued)

Table 12.1 (continued)

Competence area	Indicator	Matches with the RSP model
Communication, professional collaboration, problem-solving	7.1. Takes initiatives in collaboration with teacher colleagues, staff members and parents in order to educate students	X
	7.2. Forms and applies collaboration rules and principles for/with students, based on the pedagogical programme and documents considering students' developmental needs and stages	
	7.3. Maintains a credible and accessible communication in their activities considering the pedagogical programme and aims of the school	X
	7.4. Requires regular feedback on their pedagogical work and is open to those	X
	7.5. Elaborates notions in professional discourses, being able to change others' points of views while being able to adapt their own	X
Professional dedication and responsibility for professional learning	8.1. Aware of personal traits and able to adapt to professional roles' expectations	X
	8.2. In solving pedagogical tasks, collaborates with colleagues, their professional community, other staff members or other experts supporting the development and education of the child	X
	8.3. Participates in professional collaboration, taking initiatives with its suggestions about problems and solutions	X
	8.4. In their pedagogical work, shows a reflective approach	X
	8.5. Considers continuous professional learning important and applies newly attained knowledge efficiently in their pedagogical practice	X

Structural Questions

Competence management for teachers involves issues such as recognition of the relevance of the competence, the proof of learning and the reward of achievement (North et al. 2018). For these purposes an e-portfolio (Falus and Kimmel 2009) is created by the in-service educators during the qualification process while working with their evaluators (Kotschy 2012). This has determined the form of assessment of ESD competences, which is fortunate given that a portfolio is recognised as an appropriate tool in respect of ESD competences (Steiner et al. 2012) (see also Chap. 21).

In Hungary, one-third of educators in general education institutions are part of the whole-school ESD network (operating since 2000) using ESD quality criteria (Breiting et al. 2005), which means that they have previous experience with ESD indicators. This was a fundamental point to rely on during the introduction of the model. However, accessibility and legibility remained challenges to address. Indicators and standards needed to be formulated in a way that each educator could relate to them. In the supporting documents and training, more elaboration helped them to find evidence from their practice: 31 different guidelines were prepared for educators, and online mini-courses were created to prepare educators for the qualification process. Furthermore, ESD has been strengthened in the teacher mentoring system in regional pedagogical centres, which opens ways for future systemic development. This established a basis to transform teacher education: in 2020 and 2021, ESD competences were integrated into initial teacher training based on this work.

Conclusion

It was stated that the mere elaboration or introduction of ESD competence frameworks have not resulted in profound changes in the practice of ESD (Wals 2017). However, it was also proven by pilots that a simplified and concise ESD competence framework may imply new ways of improving educators' ESD competences (Vare et al. 2019). The question about the impact of introducing ESD competences remains open in Hungary. Nevertheless, the facts that (1) they could be well integrated into the existing competence framework of national qualifications and (2) educators and evaluators could evidently use the renewed guidance materials (as supported by successful qualifications) serves as a reference for further efforts in ESD.

References

Breiting, S., Mayer, M., & Mogensen F. (2005). *Quality criteria for ESD schools. Guidelines to enhance the quality of Education for Sustainable Development.* ENSI – SEED Network, BMBWK: Vienna

Demeter, K. (2006). *A kompetencia: Kihívások és értelmezések.* Országos Közoktatási Intézet: Budapest https://ofi.oh.gov.hu/tudastar/tanulas-tanitas. Accessed 06 Sep 2021

Falus, I. & Kimmel, M. (2009). *A Portfólió.* Gondolat: Budapest

Gugerli-Dolder, B. & Rauch, F. (2012). A model of competencies for teaching ESD – what now? Methods and experiences in acquiring competencies. In: Réti, M. & Tschapka, J. (eds) *Creating learning environments for the future Research and practice on sharing knowledge on ESD.* Environment and School Initiatives (ENSI): Kessel-Lo

Kocsis, M., Mrázik, K. & Imre, A. (2012). Megközelítési módok a tanári kompetenciák leírásához. In: Kocsis M, Sági M (ed) *Pedagógusok a pályán*, pp 61-82. Hungarian Institute of Educational Research and Development: Budapest

Korpics, M. (2015). Kompetens vagyok? A pedagógus kompetenciák megítélése a PTE-n végzett pedagógusok körében. In: Arató, F. (ed) *Horizontok II. A pedagógusképzés reformjának folytatása.* Pécsi Tudományegyetem Bölcsészettudományi Kar Neveléstudományi Intézet: Pécs http://kompetenspedagogus.hu/sites/default/files/Arato-Ferenc-szerk-Horizontok-pte-btk-ni-2015.pdf. Accessed 06 Sep 2021

Kotschy, B. (ed) (2011). *A pedagógussá válás és a szakmai fejlődés sztenderdjei.* Eszterházy Károly Főiskola: Eger

Kotschy, B. (2012). *A portfólió szerepe a pedagógusok minősítési folyamatában.* Educational Authority of Hungary: Budapest. https://www.oktatas.hu/pub_bin/unios_projektek/tamop_315_pedkepzes_fejl/3_1_5_1_hirlevel_Kotschy_Beata__portfolio.pdf. Accessed 20 Jun 2021

Kotschy, B. (2014). A pedgógiai munka értékelése – A pedagógusok minősítése. *Könyv és nevelés* 4, 16 (2014): 102-110. https://epa.oszk.hu/03300/03300/00022/pdf/EPA03300_konyv_es_neveles_2014_4_102-110.pdf. Accessed 06 Sep 2021

North, K., Reinhardt, B., Sieber-Suter, B. (2018). Kompetenzmanagement in der Praxis. (3. Auflage). Mitarbeiterkompetenzen systematisch identifizieren, nutzen und entwickeln. Mit vielen Praxisbeispielen, Springer Gabler, https://doi.org/10.1007/978-3-658-16872-8

Rauch, F., Steiner, R. (2013). *Competences for education for sustainable development in teacher education. CEPS Journal* 3 (2013) 1: 9–24

Rauch, F., Streissler, A. & Steiner, R. (2008). *Kompetenzen für Bildung für Nachhaltige Entwicklung (KOMBiNE). Konzepte und Anregungen für die Praxis.* BMUKK: Vienna

Réti, M. (2019). *Teachers' ESD competence framework in Hungary.* Workshop at the Carpathian Convention Seminar on ESD. Budapest 17 April 2019

Sleurs, W. (2008). *Competencies for ESD (Education for Sustainable Development) teachers. A framework to integrate ESD in the curriculum of teacher training institutes.* CSCT project: Brussels. https://unece.org/fileadmin/DAM/env/esd/inf.meeting.docs/EGonInd/8mtg/CSCT%20Handbook_Extract.pdf. Accessed 20 Jun 2021

Steiner, R., Sieber, B., Bäumler, E. (2012). Competencies for teaching ESD. Systematic development of ESD Competencies. The portfolio as a tool for working on individual and team competencies in schools. In: Réti, M., Tschapka, J. (ed) *Creating learning environments for the future Research and practice on sharing knowledge on ESD.* Environment and School Initiatives (ENSI): Kessel-Lo

UNECE (2011). *Learning for the future: Competences in Education for Sustainable Development.* United Nations Economic Commission for Europe (UNECE) Steering Committee on Education for Sustainable Development on 7 April 2011. Programma Leren voor Duurzame Ontwikkeling: Utrecht. https://unece.

org/fileadmin/DAM/env/esd/ESD_Publications/Competences_Publication.pdf. Accessed 20 Jun 2021

Vare, P., Arro, G., de Hamer, A., Del Gobbo, G., de Vries, G., Farioli, F., Kadji-Beltran, C., Kangur, M., Mayer, M., Millican, R., Nijdam, C., Réti, M. & Zachariou, A. (2019). Devising a Competence-Based Training Program for Educators of Sustainable Development: Lessons Learned. *Sustainability*, 11 (7): 1890

Wals, A. (2017). Can we meet the sustainability challenges? The role of education and lifelong learning. *European Journal of Education* 52 (4): 404 – 413

13 The Rise of Competence-Based Learning in the Netherlands and Its Impact on the Implementation of ESD at Marnix Academie

Gerben de Vries, Stella van der Wal-Maris, André de Hamer, and Carlien Nijdam

Abstract

In this chapter we focus on the rise of Competence-Based Learning (CBL) in The Netherlands and its impact on the implementation of education for sustainable development (ESD) in Marnix Academie (MA), Teacher Training Institute for primary education in Utrecht. We conclude that CBL did help with implementing the process, the 'how' of ESD. The content, the 'what', owned by teachers, is harder to change as it affects the autonomy and competence of teachers. Working with the institute's culture, using MA's approach of decision-making and connecting the content of ESD to its mission, proved effective. External influences, such as the Sustainable Development Goals and UNESCO membership also helped to facilitate change. We conclude that implementing ESD successfully is a slow, people-orientated, culture-focussed process that needs a multi-level, whole institution approach. Change agents need to be aware of this.

Keywords

Whole school approach · Levinas · Marnix Academie · Implementing ESD · AISHE/Prise

Introduction

After a brief explanation of some terminology, this chapter gives an overview of the rise of CBL in Dutch teacher training institutes (TTIs) before focusing on its impact on Marnix Academie Teacher Training Institute for Primary Education in Utrecht (MA). From there we move on to developments regarding ESD at Marnix Academie and the role of CBL in it. We do so by using AISHE/Prise (Roorda 2001; de Vries and de Hamer 2014) as a starting point, and the Whole School Approach (van der Meer et al. 2017; Leren voor Morgen 2020) as an instrument for analysis. We end this chapter with conclusions and lessons learned.

G. de Vries (✉) · A. de Hamer
Duurzame PABO, Sint Kruis, The Netherlands

S. van der Wal-Maris
Marnix Academie, Utrecht, The Netherlands

C. Nijdam
Hogeschool Zeeland, Vlissingen, The Netherlands

P. Vare et al. (eds.), *Competences in Education for Sustainable Development*, Sustainable Development Goals Series, https://doi.org/10.1007/978-3-030-91055-6_13

Some Terminology

Competence-Based Learning (CBL) has been discussed elsewhere in this book. Here we need to emphasise that CBL is an educational methodology, an educational approach that has no content of its own and that, according to the Glossary of Education Reform (2014), can take a wide variety of forms. It might answer questions about the *how* of education, it does not answer questions about the *what*. We shall come back to his later.

Education for Sustainable Development (ESD) has been introduced elsewhere (Chap. 1). For the understanding of this chapter we need to emphasise that it is characterised in a broad sense, as an education with a specific process, the *how* (Wals and Nolan 2012) and specific content, the *what*, based on the Sustainable Development Goals (SDGs) (Fig. 13.1).

Marnix Academie (MA) presents itself on the internet as a relatively small, open-minded, protestant-Christian University of Applied Sciences (www.hsmarnix.nl 2020); it only serves primary education. Socially, it expresses its identity in a horizontal way with great care for students, staff and interaction between people of different cultures and religions. Keywords are *competent*, *engaged* and *inspired*. Its philosophical, pedagogical and educational identity is exemplified by its emphasis on UNESCO membership, global citizenship, human rights, sustainable development and intercultural understanding. Crucial for its mission is, among others, the work of Levinas in relation to 'The Other' (Levinas 1975; Engelen 1985). Literature (de Kort 2019) and yearly reports illustrate MA's consistency in mission and identity from its founding in 1985 until the present.

Methodology

Data for this chapter have been collected from Marnix-documents and more general literature. These data have been chronologically organised. Where needed, short additional interviews were held with Marnix employees for clarification. These include three lecturers who started paying attention to sustainable development (SD) around the year 2000 and added their own knowledge early in the development stage. The current ESD and UNESCO coordinator added knowledge on recent developments about UNESCO member-

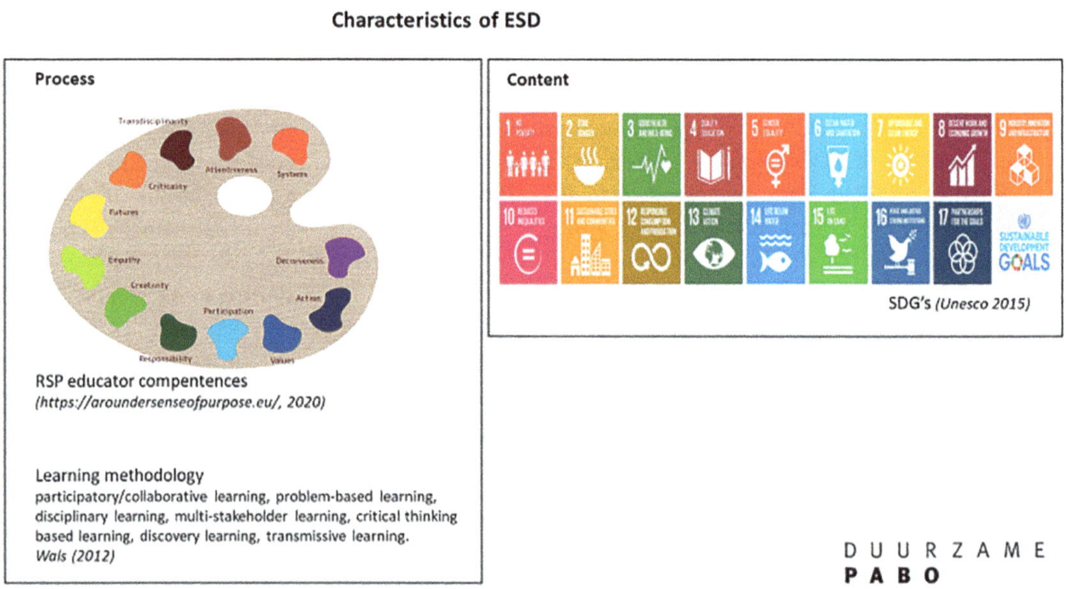

Fig. 13.1 ESD in process and content

ship and its influence on Marnix Academie. The facility manager added knowledge about matters concerning facility management and housing. The endowed professor of 'future-orientated education' added knowledge about Marnix identity, its consistency over the years and social entrepreneurship. Four lecturers, members of the taskforce for curriculum renewal and the UNESCO committee added information on recent developments. Their contributions were woven into this chapter. A concept text of this chapter was sent to them and to the vice dean, for final comments. These comments are woven into the final text, which was sent to the co-authors for approval.

Analysis of the Rise of CBL in Teacher Training Institutes in the Netherlands

In 1985, under a new law in the Netherlands, TTIs for primary education started being reshaped, becoming part of higher education, educating for teaching children aged 4–12 and taking four years instead of three. Traditionally, TTIs had their own curriculum, teaching strategies and systems of assessment with minimal Government oversight. Most TTIs had a modular system, with four semesters a year, several different lessons per day, four days of teaching and one day practice. This could lead to almost 80 different assessment points per year (320 over 4 years) not including practice assessments. Failing one of these would mean doubling some work the following semester in order to keep on track.

From 1992 quality control came under central Government and in 2003 a governmental committee ('visitatiecommissie Pabo') characterised Dutch TTIs for primary education as being broad rather than deep, lacking the theoretical level of higher education, having vague internal quality demands and disorganised assessment systems with only 9 out of 38 institutes functioning sufficiently (Sikkes 2003). As a result of this analysis, the Dutch Government took stricter control of TTIs. From 2006 this was facilitated by the implementation of CBL which included controllable quantitative and qualitative criteria backed up by law. Besides quality, the Government wanted to improve the Dutch economy (linking better teaching to a better labour force) as well as addressing a shortage of teachers (Rijksoverheid 2004).

Because of CBL, by law all TTIs in The Netherlands had to meet the same standards with freedom to add additional content. Against the background of the constitutional right of freedom of education (Rijksoverheid 2008), it is understandable that this led to huge political discussions ahead of CBL being implemented in 2006.

For most TTIs, CBL led to changes in curriculum and teaching strategies, bigger teaching units and fewer assessments. Seven competences were defined by the Foundation for Professional Quality of Teachers and Other Teaching Staff or SBL (Stichting Beroepskwaliteit leraren en ander onderwijspersoneel):

- Interpersonal competence, including indicators such as respect and personal involvement;
- Pedagogical competence, including stimulating behaviour and critical thinking;
- Professional and didactical competence, including content mastery and clear teaching;
- Organisational competence, including classroom management and planning;
- Competence in collaboration with colleagues, including asking for help and giving it;
- Competence in working with the school environment, including relationships with stakeholders and taking responsibility;
- Competence in reflection and development, including dealing with feedback and self-development.

In some indicators, characteristics of ESD appear, such as empathy, responsibility, values and collaboration.

The Government determined that from 2021 the CBL approach would be replaced with 'startbekwaamheidseisen' or starting competence demands (Rijksoverheid 2017, p. 148). These demands focus on content (especially Dutch language and Mathematics) as well as teaching methodology and pedagogy (including moral development, citizenship and social-cultural awareness). CBL is no longer compulsory, instead it is a matter of choice for each TTI. According to the Government, the new indicators are clearer, more concrete and better assessable than the seven SBL competences. Again, the aim is to improve quality control (Korthagen 2004, p. 14); however, to date these changes have not raised the quality of primary education as The Netherlands ranking in international comparisons of basic skills—language and maths—has worsened (van Nieuwstadt 2019). That said, the latest changes again show some characteristics of ESD, such as responsibility, cooperation, inspiration, moral development and critical reflection.

Marnix Academie and CBL

In 1998, as a result of internal evaluations with staff and students, MA implemented Highly Self Responsible Programming (de Kort 2019). Its aims were improved quality, better feasibility, integration of disciplines and improving students' responsibility. This meant that the step to CBL in 2006 was a small one. To express MA's mission, in addition to the seven SBL competences, an eighth competence was formulated: *Competent in inspiring, wise and value orientated doing* (van den Berg et al. 2009/2014). The eight competences together include many characteristics of ESD.

The criteria of the new law in 2017 also proved a good fit for the MA profile—and MA reacted as usual: an all-staff meeting to initiate changes in policy, curriculum and more, preparing for a new start in September 2021. These changes, described by Luijns (2019), still include Levinas and 'The Other' as basis for mission, policy and curriculum.

Analysis of the Implementation of (E)SD at MA

Piecemeal work on sustainability began in 1997 when, with the Dean's permission, three young teachers started actions to 'green the campus'. They achieved some success and the Dean signed a national covenant promising that MA would pay attention to sustainability before he retired. In 2003 his successor established a small taskforce to come up with a plan; this coincided with the development of AISHE, an auditing instrument for sustainability in higher education (Roorda 2001) so this was adopted. AISHE has twenty criteria and five levels; MA's target was set by the Dean on level three. Although action was taken, ultimately the approach failed because it was too complicated, top-down, instrumental and disruptive of the change towards CBL that was occurring at the time.

An alternative Plan B involved two components:

1. A strategy based on a review of literature concerning the implementation of ESD as a voluntary and possibly unwanted innovation. This led to a checklist (Fig. 13.2), published internally in 2012 with one of the most important messages being *work on school culture*.

2. An instrument to monitor progress: AISHE again, but now used as a model of support and development through dialogue rather than top-down control. The MA taskforce simplified it and later published it as 'Prise' (de Vries and de Hamer 2014). AISHE/Prise included much more than 'product' and 'process', showing aspects of a Whole School Approach (WSA) (van der Meer et al. 2017), as shown in Fig. 13.3.

The strategy and monitoring tool transformed MA's Plan B into something akin to a 20-board simultaneous chess game, which led to an internal report stating that '*It is unlikely that students after their studies at Marnix Academy will be teachers with an affinity for knowledge of sustainable development*' (de Vries 2004). This report resulted in new developments. The following analysis of these developments is structured in a variation of the WSA (Fig. 13.4). This includes: vision; curriculum; didactics; facility

What might help implementing ESD?

(Based on Chenoweth&Everhart (2002), Tilbury&Wortman (2004), Lozano (2006), Van Herpen (2008), Van Meer (2008), Hargreaves&Shirley (2009), Roorda (2010), Tilbury (2011).

Appoint a team and coordinator
Facilitate openly
Connect to the shared mission, vision
Connect to a shared aim
Legitimate openly
For deep change, focus on school culture
Incremental change work better than radical change
Involve students and the school network
Use a structured multiplier effect
People differ
Learn from the inside and the outside
Involve as many stakeholders as possible
Make participation in processes part of the dominant culture
Share your success, internally and outside
Use external, governmental policy as back support
Sail on the wings of change
Evaluate on a regular base, using a chosen design model
Share and discuss the outcome of the evaluations
Use the outcome for further policy development

DUURZAME PABO

Fig. 13.2 Strategy list for implementing ESD (de Vries 2012)

management; professional development (including Human Resource Management); community.

Vision

For MA, as a strongly mission- and values-driven organisation, connecting ESD to that mission was crucial for acceptance. MA's vision, focussing on Levinas and the other (van den Berg et al. 2009/2014) was summarised into MA's definition of ESD: '*awareness of, respect for and taking responsibility for yourself, The Other and the other, here and there, in past, present and future*' (de Vries 2014). This connection is still valid (van der Wal-Maris 2019).

Curriculum

ESD, based on the research of 2004, found its place within the competences in 2006 with one word: sustainability, as part of the professional and didactic competence description. ESD content at the time was provided by the eight Millennium Development Goals (MDGs) and by 'Vensters op de Wereld' (Beneker et al. 2009). Its eight SD themes were used on an annual basis as 'theme of the year', helping teachers to turn the mission statement into concrete teaching that influenced the curriculum. In 2017, the Sustainable Development Goals (SDGs) took over with changes in society also having their influence such as working with refugee children and dealing with diversity. Being a UNESCO school from 2017 is reflected in programmes, research and postgraduate courses, as shown in reports and emphasised by the UNESCO coordinator. Most notable here is 'social entrepreneurship at MA' (Marnix Academie 2020a, b, c, d), which reflects the MA definition of ESD, mentions the SDGs, has methodologies that resemble Wals and Nolan (2012) list and competences that resemble A Rounder Sense of Purpose (RSP)

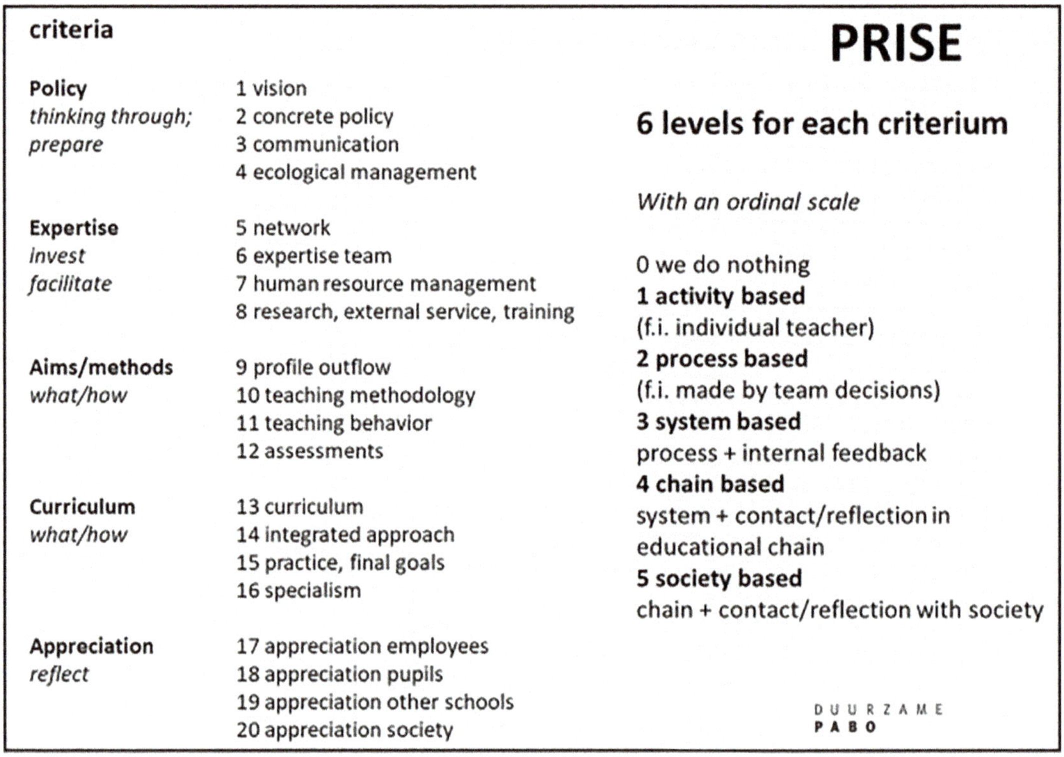

Fig. 13.3 Indicators for progress: prise criteria, based on AISHE (Roorda 2001)

Fig. 13.4 Whole school approach (van der Meer et al. 2017, as adapted by Leren voor Morgen 2020)

(Vare et al. 2019). It describes a 'LEV' learning line, in which 'LEV' stands for Levinas, Guts (Dutch; 'lef') and 'heart' in Hebrew. This will be part of the curriculum from September 2021.

In 2018, MA appointed two professors with designated assignments (www.marnixacademie.nl 2020): 'value-based leadership' and 'future-orientated education', enabling students to take their responsibility in co-creating a more righteous, democratic and meaningful way of life (van der Wal-Maris 2019). Their work is action-oriented and closely related to the 'LEV' line. This is not to suggest that things have always worked positively. In 2007 the Al Gore film, *An Inconvenient Truth,* was shown in the largest lecture hall to only three students and two teachers. Student involvement proved hard to achieve. Attempts to put on postgraduate courses in 'green' sustainable development failed because of a lack of interest from primary schools. Conversely, the social side of ESD, such as peace education or working with refugee children is strongly supported.

Teaching

Thanks to SBL, the eighth competence and MA policy, teaching in 2006 already met many of dimensions of ESD with subsequent developments such as a methodology for teaching ESD (de Vries and de Hamer 2010). The national changes proposed in 2017, to be implemented in 2021, represent the next step, yet MA does not view this as a way of achieving ESD; it simply fits into the ideology.

Facility Management

Changes towards CBL and self-steering brought about changes in buildings and facilities such as more smaller meeting rooms and individual working places. MA has also paid attention to eco-management for some years (de Vries 2004). According to the facility manager, these activities became more systematic and efficient over time. Curriculum changes, a growing student population and, more recently, COVID-19 have all served to open discussions about achieving even more environmentally friendly ways of working as well as cheaper housing.

Human Resource Management (HRM)

Changes to CBL and ESD were prepared the MA way: democratic, bottom-up, involving all staff, from dean to cleaning personnel, as respectfully as possible (de Kort 2019), in order to improve acceptance and support. Changes were not always greeted enthusiastically and sometimes led to frustration, particularly where it affected staff autonomy and competence. Careful HRM, retirement and job changes helped as ESD and concern for world citizenship became criteria in new appointments. While the ESD taskforce was reduced to one member with only 40 h a year due to diminished management support, progress is facilitated by a Plan C: implementing (E)SD through a process of constant dialogue with all colleagues, thereby contributing to a broad acceptance. Rethinking strategies for securing acceptance became important; this led to the UNESCO membership while all successes, large or small, were supported and communicated internally making all staff part of the success. Success has also been communicated explicitly to the outside world, which has built MA's reputation as an ESD-TTI.

Professional Development

Professional development at MA is organised through all-staff meetings almost every 5 years plus smaller events for more specific aims. Furthermore, each employee has a personal development plan. ESD slowly became part of this agenda. Most teachers were not interested in 2003. All kinds of activities were organised, working on change of culture; the UNESCO membership helped while wider societal changes (e.g., the Paris agreement of 2015) also raised awareness.

ESD-related professional development at MA was monitored from 2003 using AISHE/Prise. Imposing this approach failed in 2003 but success was achieved by aligning ESD with the description of the SBL competences. AISHE2, including a special governmental tribute, was welcomed in 2008. AISHE3 was reached in 2012, which made MA officially the highest ranking ESD-TTI in the country (Marnix Academie 2013). Each AISHE audit led to a report that was discussed and became a reason for change. In 2017, MA left the AISHE system and became a UNESCO school (Marnix Academie 2018).

There were several reasons for change: MA policy moved towards world citizenship, UNESCO was more appealing, fitted better into MA's mission, was less competitive and was not viewed as a tool of management.

Community

Here, Duurzame PABO, an NGO that promotes networks of TTIs for pre-school and primary schools that are actively involved in ESD proved to be crucial. It was the main connection to organisations and meetings, providing MA with contacts such as guest speakers and keynotes. Eventually it was this connection that led to the UNESCO membership. During a visit to the TEESnet conference in 2017, with almost all science and humanities teachers, the SDGs were introduced and became part of MA's curriculum ever since. The UNECE competences (de Hamer and Leussink 2012) also became part of the curriculum, as did the RSP competences in 2017.

Governmental policy on CBL also helped as do new general aims for education in The Netherlands (Curriculum.nu 2020). This might give a boost to implementation of ESD into primary schools and TTIs and makes implementation easier than was the case 20 years ago. In the coming years MA will work on dealing with diversity by using its network of over 350 primary schools, including inner-city schools.

Conclusions and Lessons Learned: Did CBL Help with Implementing ESD?

At first sight, in 2003 it did not help; CBL was being implemented at MA and implementing ESD was seen as a threat to the CBL process. However, in 2006, with the official start of CBL, it did help. Not so much because of the CBL system, but because of the way the government described the indicators and the way MA added an extra competence with additional indicators, filling it with its mission. Changes in MA's competence profile in 2009 and 2014 further strengthened the implementation of the process, the *how* of ESD.

The content, the *what*, owned by teachers, was harder to change as it affected the autonomy and competence of teachers even further than the SBL competences and ESD was not forced by the government in the same way. Working on the school culture, using MA's ways of decision-making and connecting the content of ESD to the mission, the concept of The Other, proved successful, supported along the way by the SDGs and the UNESCO membership.

In 2017, preparing for implementation in 2021, the two came together. ESD finally became a mainstream part of MA. Again, the change was forced by law and filled by MA in its own way—with LEV (Levinas, guts and heart). To make this all possible, a clear goal (AISHE3, later UNESCO membership) and a clear strategy were helpful. An almost constant effort to influence the culture, in a multi-level process, working both bottom-up and top-down and adjusting to the MA way of working slowly made change happen. Connecting ESD to MA's mission proved to be crucial, something which may be unique among Dutch TTIs.

Despite the single-case character of this study, we might distinguish some conclusions that can be transferable to other situations:

- Implementing ESD can be seen as a non-compulsory, unwanted change; it needs to be approached in this way. Do not expect it to be fun unless you make it fun.

- External pressures, such as laws, can make things happen but these need a careful translation to fit in the school system and culture.
- Implementing ESD successfully is a slow, people-orientated, culture-focussed process that needs a multi-level whole school approach.
- Change agents/ESD innovators should be aware of the need for positivity, stubbornness, wisdom, perseverance and a sense of strategy, keeping their eyes on the prize over the long term; they might learn how to play chess first.

CBL can be of help in implementing ESD, although other approaches might fit even better. Almost a century ago and unknowingly, Parkhurst, Freinet, Petersen and Boeke were already on the move (see Ahlers 1982). It might be interesting to see what we can learn from these reform educationalists.

Acknowledgements This chapter has been written with support of many employees of MA. Thanks to all of you, especially Caroline, Patricia and Rob, who dared to start the fire! One of them ended an interview for this chapter with: 'I am so very proud of what MA achieved in this field'.

References

Ahlers, J. (1982). *Wat is dat voor een school*, Deventer, Netherlands, Van Loghum Slaterus.

Beneker, T., van Stalborch, M. & van der Vaart, R. (2009). *Vensters op de Wereld,* Utrecht, Netherlands: NCDO.

Chenoweth, T.G. & Everhart, R.B. (2002) *Navigating comprehensive school change*, Larchmont, USA: Eye on Education.

Curriculum.nu (2020). *Bouwstenen voor een herzien curriculum,* www.curriculum.nu/voorstellen/.

de Hamer, A. & Leussink, E. (2012) *Leerkrachtcompetenties duurzaamheid*, Netherlands, Nieuwkoop: Ecodrukkers.

de Kort, B. (2019). *Identiteitsontwikkeling van en in een lerarenopleiding*, Utrecht, Netherlands: Eburon.

de Vries, G. (2004). *0-meting duurzaamheid en Marnix Academie*, internal document, Utrecht, Netherlands: Marnix Academie.

de Vries, G. (2012). *Advies Bestuur Kopwerk*, internal document, Utrecht, Netherlands: Marnix Academie.

de Vries, G. (2014). *Mission statement Duurzame Ontwikkeling*, internal document, Utrecht, Netherlands, Marnix Academie.

de Vries, G. & de Hamer, A. (2010). *Naar een didactisch model voor duurzame ontwikkeling,* in Praxis nr 5, 's Hertogenbosch, Netherlands: Malmberg.

de Vries, G. & de Hamer, A. (2014). *Prise, tool for assessment and evaluation in ESD*, Learning Teacher Magazine 3-4, Karlstad, Sweden: LTN.

Educating with a Rounder Sense of Purpose (2020). https://aroundersenseofpurpose.eu/.

Engelen, J.C.M. (1985) *Het gelaat, jij die mij aanziet*, Hilversum, Netherlands: Gooi en Sticht.

Glossary of Education Reform (2014, May 14). *Competency-Based learning* www.edglossary.org/competency-based-learning.

Hargreaves, A., & Shirley D. (2009). *The Fourth Way*, Corwin Thousand Oaks, USA 2009.

Korthagen, F, (2004). *Zin en onzin van competentiegericht opleiden*, Velon tijdschrift voor lerarenopleiders 2004/25-1

Leren voor Morgen (2020). *Update WSA October 2020*, internal document, Utrecht.

Levinas, E. (1975). *Het menselijk gelaat,* Bilthoven, Netherlands: Ambo.

Lozano, R. (2006). *Incorporation and institutionalization of SD;* Journey of Cleaner Production 14 (2006) 787-796, www.elsevier.com.

Luijns, K., (2019). *Afstemming 2020,* Utrecht, Netherlands: Marnix Academie.

Marnix Academie (2006). *De Marnix Academie in 2005, jaarverslag*, Utrecht, Netherlands: Marnix Academie.

Marnix Academie (2013). *De Marnix Academie in 2012, jaarverslag*, Utrecht, Netherlands: Marnix Academie.

Marnix Academie (2018). *De Marnix Academie in 2017, jaarverslag*, Utrecht, Netherlands: Marnix Academie.

Marnix Academie (2020a). *About the Marnix Academie.* https://www.marnixacademie.nl/English.

Marnix Academie (2020b). *De Marnix Academie in 2019, jaarverslag*, Utrecht, Netherlands: Marnix Academie.

Marnix Academie (2020c) *Lectoraten,* www.marnixacademie.nl/onderzoek/lectoraten.

Marnix Academie (2020d). *Poster Sociaal Ondernemen*, Utrecht, Netherlands: Marnix Academie.

Rijksoverheid (2004). *Hoger Onderwijs en Onderzoeksplan 2004*, Den Haag, Netherlands: Min. OCW.

Rijksoverheid (2008). The constitution of the Kingdom of The Netherlands, Den Haag, Nederland, Min. of the interior.

Rijksoverheid (2017). *Wijziging besluit bekwaamheidseisen onderwijspersoneel*, Staatsblad van het Koninkrijk der Nederlanden, 2017-148, Den Haag, Netherlands: Min. OCW.

Roorda, N. (2001). *Aishe, instrument for sustainability in higher education*, Amsterdam, Netherlands: DHO.

Roorda, N. (2010). *Sailing on the wings of change,* Maastricht, Netherlands: UPM.

Sikkes, R., (2003). *De onmogelijke opdracht van de pabo*, Aob 17, 4-10-2003

Tilbury, D. (2011). *Education for Sustainable Development,* Paris, France: UNESCO

Tilbury, D. & Wortman, D. (2004). *Engaging people in sustainability,* Gland, Switzerland: IUCN.

UNESCO (2015). *Sustainable Development Goals*, www.unesco.org/sustainabledevelopmentgoals.

van den Berg, B., de Kort, B., & van der Wal-Maris, S. (2009/2014). *Bekwaam, betrokken en bevlogen,* Utrecht, Netherlands, Marnix Academie.

van der Meer, P. (2008). *Identiteit, levensbeschouwing en duurzame ontwikkeling*, in Jansen, e.a. 'Naar een duurzame pabo', Nijkerk, Netherlands: Callenbach.

van der Meer, S., Swart, H., Wals, A., Wesselink, R. & Frijters (2017). *The whole School Approach to Sustainability and System change,* Wageningen, Netherlands: Wageningen University.

van der Wal-Maris, S. (2019). *In Verbinding, onderwijs met het oog op de toekomst*, Utrecht, Netherlands: Marnix Academie.

van Herpen, M. (2008). *Duurzaam opvoeden en ontwikkelen,* Antwerpen, Belgium: Garant.

van Nieuwstadt, M. (2019). *Nederlandse Pisa Scores zakken weg*, https://www.aob.nl 3-12-19.

Vare, P., Arro, G., de Hamer, A., Del Gobbo, G., de Vries, G., Farioli, F., Kadji-Beltran, C.; Kangur, M., Mayer, M., Millican, R., Nijdam, C., Réti, M., Zachariou, A. (2019). Devising a Competence-Based Training Program for Educators of Sustainable Development: Lessons Learned. *Sustainability* 2019, 11, 1890. https://doi.org/10.3390/su11071890

Wals, A., & Nolan, C. (2012). *Shaping the Education of Tomorrow,* Paris, France: UNESCO.

14 Working on ESD Teacher Competences in Switzerland: A Case Study at the University of Teacher Education Vaud

Alain Pache and Nadia Lausselet

Abstract

After a short introduction to the Swiss context, this chapter presents the analysis of a pre-service training course aimed at developing ESD competences. Student learning is identified, as well as the strengths and limitations of the competence model used. By linking these findings to the institution's action plan, two useful avenues for training are identified: on the one hand, making the whole institution approach more explicit in order to allow students to experience being change agents; on the other, allowing more frequent links between ESD approaches and the training courses themselves. The whole process requires a lot of internal lobbying in order to achieve strong sustainability objectives.

Keywords

ESD · Competences · Primary school · Teacher training · Whole school approach

A. Pache (✉) · N. Lausselet
University of Teacher Education State of Vaud, Lausanne, Switzerland
e-mail: alain.pache@hepl.ch

Introduction: A Supportive Context for Implementing ESD

Sustainable development (SD) and education for sustainable development (ESD) have found their way in numerous official texts in Switzerland. However, although the context seems to support implementing ESD, Curnier (2017) has shown that in French-speaking Switzerland, intentions in favor of ESD contained in international and national official texts become diluted the closer one gets to the operational level (i.e., the "canton" or state level), and don't find their way into cantonal educational steering tools.

One of these French-speaking cantons, namely Vaud, has tackled this issue: a new head of the Department of Education, Youth and Culture has declared SD and ESD, in a whole institution perspective, as a priority and has established a related working group. Its aim is to turn the local school system into a "model of sustainability."[1] It is within this dynamic that the University of Teacher Education Vaud, hereafter named HEP Vaud, has opened new courses working on ESD competences along the "Rounder Sense of Purpose" model (RSP, see Chap. 5). At the same time, the university has initiated an internal

[1] https://ecolevaudoisedurable.ch/vision-et-durabilite.

P. Vare et al. (eds.), *Competences in Education for Sustainable Development*, Sustainable Development Goals Series, https://doi.org/10.1007/978-3-030-91055-6_14

action plan in the field of SD and ESD. The underlying logic echoes the idea that educators should act as change agents in a process of organizational learning while being able to design supportive learning environments for ESD (UNESCO 2017).

This chapter looks at this double approach and possible dynamics within it. A first section will clarify the conceptual framework in which this double approach is anchored. A second section will look at students' perspective on a new course for pre-service primary teachers aimed at developing ESD competences along the RSP model. A third section will briefly look at the parallel work done on organizational development processes that have been launched at HEP Vaud. Finally, a fourth section will discuss two main points emerging from the combined analysis of these two approaches before a conclusion offers possible perspectives.

Conceptual Framework: A Call for a Paradigm Shift

A literature review by Curnier (2017) highlights the need to transform society by moving away from the paradigm of Modernity, i.e., a relationship to the world built in the West over the past centuries. This would involve transforming schools by setting them on the path of a socio-ecological transition based on the principles of strong sustainability, in which the author points out, human activities are placed in an ecological context that represents a "binding framework" (p. 81), and that the economy's function is to meet basic human needs and thus contribute to a better society. As for the socio-ecological transition, it refers to Rockström et al. (2009) idea of "great transitions" seen as a possible scenario for the future, which implies a societal paradigm shift in order to meet the challenge "of allowing human activities to reproduce in the long term within the impassable limits imposed by the functioning of the Biosphere" (Curnier 2017, p. 100).

The prescribed curriculum Curnier proposes is based on a "transformative" posture (Sterling 2011), which seeks to question existing structures and paradigms in order to identify their deep roots and make them evolve through a triple cultural, historiographical, and artistic rupture (p. 123). Such a posture requires a redefinition of the aims of education and its values, the establishment of a new relationship to the world and to others, as well as a new relationship to knowledge. The new curriculum he proposes would therefore be based on the development of competences (critical thinking, complex thinking, prospective thinking, ethics, taking action, and so forth) as well as on the central concepts of Anthropocene, Biosphere, Development, and Socio-ecological Systems. The work would be organized along interdisciplinary teaching sequences (for some carried out outdoors) and would therefore imply a renewal of school habits (Vincent 1994).

Based on this conception of ESD, and using the RSP model as a framework, a new course for pre-service primary teachers has opened, to which we now turn.

The Example of a Course for Pre-service Primary School Teachers

As part of the RSP II research project (see Chap. 5), we have set up an experimental course in the spring semesters 2019 and 2020. This three ECTS credit course, entitled "Practicing social sciences to learn how to read the world: ESD, ethics and geography" is aimed at upcoming primary teachers in the second year of their Bachelor's degree.[2] Eight three-hour sessions were spread over the semester: The twelve competences of the RSP model are worked on during the first six sessions, whereas session seven happens outdoors and aims at integrating the previously seen elements. Finally, session eight serves as a formative assessment before the exams (Fig. 14.1).

[2] 125 students chose this module in 2019 and 110 chose it in 2020. This corresponds overall to 60% of the total number of second-year students. The other students worked on a module entitled "History and Citizenship."

Session 1	Session 2	Session 3	Session 4	Session 5	Session 6	Session 7	Session 8
Systems Attentiveness	Transdisciplinarity Criticality	Futures Empathy	Creativity Responsibility	Participation Engagement	Action Decisiveness	Outdoor Education	Formative Assessment

Fig. 14.1 The course's architecture

In each session, links are made between RSP competences, the SDGs, and formal teaching resources. This demonstrates the need to work on content that is interdisciplinary in nature as well as the disciplinary contributions (in geography and in ethics).

At the end of the semester, the students submitted a portfolio in which they had to attest to which competences were acquired and how they could be worked on with pupils. In addition, a focus group was organized at the end of each semester with volunteer students (four students in 2019 and one student in 2020[3]). These interviews were transcribed and both sets of data were analyzed on the basis of the following research questions:

- What are the competences you have developed in this module?
- What mechanisms do you think would enable these competences to be assessed?
- What are the strengths and limitations of the RSP competence framework?

The results are presented hereafter following this structure.

What Students Have Learned

What mostly struck students was that it was possible to work on a sustainability issue in an interdisciplinary way. This is not surprising as interdisciplinarity is worked on only in the students' third study-year, which takes place after this course. Here is what Julie says about it (focus group 2019):

> So for me the great revelation was the, the interdisciplinarity precisely because for me sustainable development is a subject that is fascinating and that fascinates me and for all that, uh, I didn't see in the timetable well before this course how to, the, where to put it, where to do it in such a way, so that there is a meaning, not just in little bits and pieces ... and so it was really interesting to see that it touches all the subjects and that all the subjects can be used to convey messages around that, especially the critical sense, which is very important to me personally, given what's happening in the world at the moment, … what I learned by looking at the theory is that there must be a recurrence in the teaching of how to exercise one's critical sense ... in fact we can do sustainable development in French, we can do sustainable development in history, in many other subjects and... I didn't necessarily have this awareness before, so that's what I remember.

Other students mentioned elements that were more at the level of sustainability-related concepts helping them to frame their professional thoughts (e.g., strong versus weak sustainability, the Anthropocene, the cradle-to-cradle model), or examples seen such as the "Just Community" schools.

When it comes to identifying the competences developed within the course however, students are more circumspect. They mention that they have come a long way, but that there is still a lot of work to be done. Christiane, for example, reports that she would have liked to spend more time on it:

> I don't feel that I master the twelve competences thoroughly. I have the impression that I've scratched the surface of everything, I know more or less what it's all about, but there are so many components when you really take it seriously, even with the help of the RSP site, well it's so rich that I don't pretend here that I am trained, I think that, we could add another course...

Another student, Axelle, points out the different time frames required in the acquisition of

[3] We thought we would mobilize several students in 2020, but the context of the pandemic and the resulting overload resulted in many defections.

these competences with the competences related to "thinking holistically" being acquired quicker than those related to "achieving transformation" with pupils. She also makes a link between the individual and collective scales:

> I think the ones I find the most difficult is to achieve a transformation because thinking holistically I think it's, in itself I don't think it's much of a problem, but to envision change and to achieve (support) a transformation, it's something you have to see at the level of, at the level of the continent, at the level of the planet, and that's something that is very, very complex.

However, one of her colleagues reacted specifying that the transformation must above all be carried out at the individual level and not at the continental level, which implies that both students interpret this point of the RSP framework in their own way.

How to Assess Competences

With regard to the mechanisms for assessing these competences, the students identified four possible domains that have to be mastered so as to show that a competence has been developed. According to them, the students have to:

- use the right terms when talking about SD
- master certain tools that help implement ESD (for example the concept map to prepare a complex theme)
- master certain methods that are relevant to ESD (for example, the inquiry method, the sustainable prospective narrative method, or the debate method)
- be able to develop self-evaluation grids for pupils

Strengths and Limitations of the RSP Model

An examination of the portfolios submitted at the end of the semester allowed us to highlight two strengths and define three limits of the RSP model.

The overall feedback on the model was positive. The first strength that was highlighted is that the model gives a good overview of the set of ESD-related competences and is considered attractive and handy to use, especially thanks to the competence-related participatory activities. Furthermore, students appreciated the fact that the model allows them to clearly identify ESD-relevant knowledge.

With regard to the limitations, the students highlighted the difficulty of combining and articulating various RSP competences in teaching-learning situations. In fact, the model proposes twelve competences presented as a grid[4] rather than in an articulated manner. Moreover, each cell of the model contains underlying components (UCs), which mention what has to be acquired within the broader competence (e.g., *Systems*), as if it related only to this one competence. "Understanding and criticizing different models of sustainable development" (UC1.1b mentioned under Systems[5]), for example, does not only relate to the Systems competence but also to the competences *Attentiveness*, *Futures*, and *Values*. Working on exemplified teaching situations and relating them with several articulated competences would therefore be useful in order to show the linkage with competences including the dimension of content knowledge.

The second point is about the links between competences and SD issues. According to our students, the RSP model does not insist enough on the fact that complex thinking, critical thinking, prospective thinking, or creative thinking can be worked on without reference to sustainability issues and, in particular, to planetary boundaries as defined by Rockström et al. (2009). Such an approach could, in our view, be counterproductive as it would not focus on current priorities. In other words, although the model is framed within SD, it seems important to make this explicit at all stages as prospective thinking, if decontextualized, can also be worked on in a neo-

[4] At the time, only the grid version of the model had been translated in French, and students were working with this, not with the palette.

[5] https://aroundersenseofpurpose.eu/framework/th-int.

liberal perspective (Lausselet 2019, 2020; Corres et al. 2020).

Finally, the third point highlights the fact that the model does not really distinguish between teaching competences and the competences to be acquired by pupils, which is considered confusing. This echoes authors such as Bertschy et al. (2013) that state that it is meaningful to make the specificities of ESD teacher competences explicit and to link them with more general and broadly acknowledged teacher competence models such as the one of Baumert and Kunter (2006).

Having looked at issues related to implementing an ESD teacher competences model, we now focus on the second change area, the launch of an institutional action plan.

An Action Plan for HEP Vaud

The action plan accepted in January 2020 by the HEP Vaud Board of Directors is structured along three axes: Institution, Training, and Research and development. It has three objectives:

- The HEP integrates the idea of sustainability into its operations and is recognized for its innovative approaches in the field.
- The HEP's training offer enables future teachers to implement ESD in a professional way.
- The HEP is recognized as a significant actor by its contribution to the scientific discourse on ESD.

To implement this action plan, a working group, bringing together a total of twenty-eight colleagues, has been set up. It consists of a core group, an operational group, and an advisory group. The members of the core group belong to and coordinates the other two groups. Finally, the fourth member of the core group coordinates the whole working group (Fig. 14.2).

For the first year, a focus has been set on finding a common language and on integrating colleagues in a participatory process looking at possible ways forward. A range of internal actions (sustainability lunches, conferences, workshops, revision of curricula) have been undertaken, new training offers have been designed (e.g., CAS for teachers, DAS module for heads of school), and scientific workshops[6] have been organized.

Time has not allowed the systematic collection and analysis of empirical data here but some general conclusions can be drawn after a first year: a collective dynamic has been successfully launched, with visible results; a new position for a person in charge of sustainability within the institution has been opened; new courses are offered; new research projects have been launched; various measures have been taken at an institutional level (e.g., the food offered in the cafeteria). However, the three axes still evolve quite separately and no common definition, be it of SD or of ESD, has yet been adopted. This confusion results in partially unsatisfactory approaches and "greenwashing" tendencies. Moreover, the structures of the institution and the underlying mental models in play are not yet consistent with a transformative approach.

Discussion

Having looked at students' perspectives regarding a course based on the RSP framework and offered a brief review of the beginning of an ESD related process within the institution, two main elements are now being discussed among involved stakeholders.

Making a whole institution approach more explicit and taking it as a vision that helps orientate all actions seems a necessity. A whole institution approach would help to address the difficulties mentioned by students while working with the RSP framework. Students could experience being change agents and work more concretely on those competences within the field of "achieving transformation" that appear to be more difficult to grasp. They could address the

[6]An overview of the realized events can be seen on the website of our laboratory (unfortunately not yet translated into English!): http://www.hepl.ch/cms/accueil/recherche/laboratoires-hep-vaud/liredd/actualites-et-evenements.html.

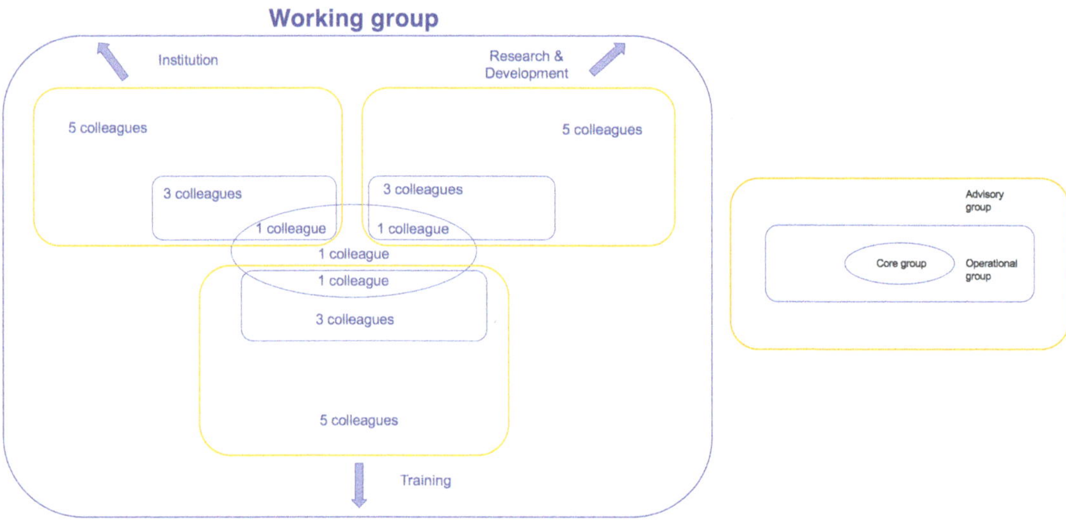

Fig. 14.2 The composition and organization of the sustainability working group

tension between individual transformation and institutional (thus structural) transformation, and apprehend possible ways of working similarly with pupils. They would also have the opportunity to articulate various ESD competences in real-life situations and work more closely on the dynamics between interdisciplinary situations and disciplinary approaches. In brief, it would facilitate the creation of a supportive ecology of learning (Wals 2020) for developing ESD competences within the conceptual framework presented above.

In order to address the fact that it is difficult to work on the twelve competences of the RSP framework other than superficially within a single course, and that some competences require more time to be developed, a closer look has been given on how to articulate various new and existing courses that can contribute to progressively address these competences, resulting in the structure presented in Fig. 14.3.

The first course on "Concepts on and for SD and ESD," which is new, starts this year for around 480 students. A coaching offer for the teacher educators in charge of courses contributing to ESD in year two has been set up, in order to help them make the contribution to ESD clear. Several colleagues have opened up new compulsory elective courses in year 3. The main resistance is likely to be encountered regarding the interdisciplinary study week because neither time nor credits are available within the study plan but the suggestion is considered meaningful and the issue is being discussed. The idea is that this study week will link all previous contributions and train participants in ways to implement them in class.

Perspectives

The case study described in this chapter shows the connection between a supportive political context and an iterative dynamic at the institutional level between work on ESD competences and an organizational learning process. This dialogic relationship between an institutional process tending towards a supportive learning environment and a pedagogical process structured around ESD competences could benefit from further studies as it has the potential to result in organically developing ecologies of learning and making a whole institution approach become a reality. Ways of implementing this in a learning progression, using existing courses that can be adjusted to contribute to ESD and complementing this with new courses oriented specifically to ESD is another area worth exploring. The RSP model can be used as an interesting framework providing common ground when articulat-

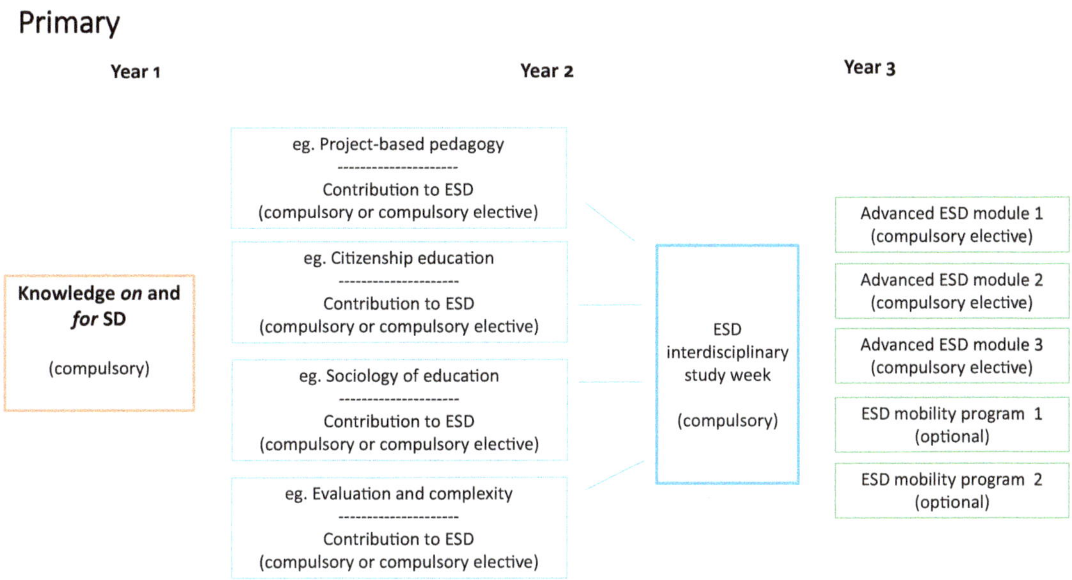

Fig. 14.3 A curricular perspective in order to develop ESD competences in teacher education: the example for primary pre-service teachers

ing the contribution of various ESD courses but it has to be implemented while keeping in mind possible difficulties such as those mentioned in this chapter. The whole process requires a great deal of internal lobbying work so that the various stakeholders can be convinced that it is absolutely essential to train in new competences, not only to deal with present and upcoming challenges, but above all to plan a world that is fairer and takes into account the planetary boundaries.

References

Baumann, S., Lausselet, N. & Pache, A. (2019). *L'EDD dans la formation des enseignant.e.s. Etat des lieux-juillet 2019*. Bern : swissuniversities.

Baumert, J. & Kunter, M. (2006). Stichwort: Professionelle Kompetenz von Lehrkräften. *Zeitschrift für Erzihungswissenschaft*, 9 (4), 469-520.

Bertschy, F., Künzli David, Ch., Lehmann, M. (2013). Teachers`Competencies for the Implementation of Educational Offers in the Field of Education for Sustainable Development. *Sustainablity*, 5, 5067–5080.

Corres, A., Rieckmann, M., Espasa, A. & Ruiz-Mallén, I. (2020). Educator Competences in Sustainability Education: A Systematic Review of Frameworks. *Sustainability*, *12*(23), 9858.

Curnier, D. (2017). *Quel rôle pour l'école dans la transition écologique ?* Thèse de doctorat. Université de Lausanne : Faculté des géosciences et de l'environnement.

Lausselet, N. (2019). Im Zeitalter der SDGs unterrichten: Knackpunkte, Denkanstösse und Beispiele. Oral communication, Goethe Universität Frankfurt.

Lausselet, N. (2020). Nachaltige Entwicklung und Bildung für Nachhaltige Entwicklung als Rahmen für die Lehrerinnen-und Lerhrerbildung : Chancen, Herausforderungen und offene Fragen. Oral communication, PH Bern.

Rockström, J., Steffen, W., Noone, K. et al. (2009). A safe operating space for humanity. *Nature* 461, 472-475. https://doi.org/10.1038/461472a.

Sterling, S. (2011). Transformating Learning and Sustainability: sketching the conceptual ground. *Learning and Teaching in Higher Education* (5), 17–33.

UNESCO – United Nations Educational, Scientific and Cultural Organization (2017). *Education for Sustainable Development Goals: learning objectives*. UNESCO. https://unesdoc.unesco.org/ark:/48223/pf0000247444_eng

Vincent, G. (Ed.) (1994). *L'éducation prisonnière de la forme scolaire. Scolarisation et socialisation dans les sociétés industrielles*. Lyon: Presses universitaires de Lyon.

Wals, A.E.J. (2020). *Sustainability-oriented ecologies of learning. A response to systemic global dysfunction*. In: R. Barnett & N. Jackson (Eds.). Ecologies for Learning and Practice. Emerging Ideas, Sightings and Possibilities. Routlege: New York (pp. 61-78).

15 Development and Assessment of ESD Competences: Staff Training at the University of Vechta

Lukas Scherak and Marco Rieckmann

Abstract

This chapter presents the implementation of an ESD academic staff training programme at the University of Vechta (Germany) with a particular focus on the development and assessment of ESD competences. It reports the results of a focus group with staff training participants, which discussed the ESD competence framework developed within the Erasmus+ project *A Rounder Sense of Purpose*. The aims of the focus group discussion were twofold: to reflect on the comprehensiveness of the framework and to consider how to assess the competences concerned. This chapter thus provides empirical data to identify the further steps required to adopt this approach to staff training and the strengths and weaknesses of the RSP competence framework. It first makes clear that an integrative approach is crucial when considering the ESD competences, and then goes on to highlight the challenge of raising enthusiasm for ESD amongst teachers who do not yet have an affinity for ESD. It also raises the question of how to go beyond short workshops and facilitate learning processes that contribute to the more comprehensive development of ESD competences. Finally, it clarifies the need for methods of assessing students' competence development.

Keywords

Education for sustainable development · Educators' competences · Assessment · Staff training

Introduction

When it comes to the implementation of competence-based education for sustainable development (ESD) (Rieckmann 2018), the first question that needs to be addressed is *which competences* should be promoted (Brundiers et al. 2021; see also Chaps. 3 and 4). Secondly, the question arises of how these competences can be *developed* (Lozano and Barreiro-Gen 2021; Chap. 17); and thirdly, the question of how to *assess* the development of competences (Redman et al. 2021; Chap. 21). The *Rounder Sense of Purpose* (RSP) competence framework (Vare et al. 2019; Chap. 5)[1] provides answers to the first and in part the second question. The topic of assessment, however, was not a primary focus of the RSP project. Going beyond the project, it is

L. Scherak (✉) · M. Rieckmann
University of Vechta, Vechta, Germany
e-mail: Lukas.Scherak@uni-vechta.de

[1] https://aroundersenseofpurpose.eu/framework/themodel/.

P. Vare et al. (eds.), *Competences in Education for Sustainable Development*, Sustainable Development Goals Series, https://doi.org/10.1007/978-3-030-91055-6_15

notable that in contrast to the intensive debate on the concepts of sustainability competencies, to date ESD research has made only a few attempts to develop instruments for assessing competence development (Redman et al. 2021). Within the framework of the RSP project, an ESD academic staff training programme was introduced at the University of Vechta (Germany), focusing on the RSP competence framework in its content and structure. The implementation of this programme and the associated evaluation have delivered findings on all three of the above questions.

This chapter takes as its basis the UNESCO definition of competences: "Competencies describe the specific attributes individuals need for action and self-organisation in various complex contexts and situations. They include cognitive, affective, volitional and motivational elements; hence they are an interplay of knowledge, capacities and skills, motives and affective dispositions" (UNESCO 2017, p. 10). Competences cannot be taught to learners, rather learners develop competences when they apply their knowledge and take action. In addition to action, an important factor for competence development is reflection on the resulting experience (Weinert 2001).

The chapter first discusses the relevance of academic staff training in higher education in Germany and the basics of the development and assessment of ESD competences and then describes the introduction of the ESD staff training programme at the University of Vechta and the results of a focus group discussion with training participants. Finally, the results are discussed and conclusions drawn.

Relevance of Academic Staff Training in German Higher Education

The need for professionalisation of university teaching is nowadays widely recognised in Germany. Whereas before the introduction of the Bologna study system, the professionalisation of university teaching still provoked fierce resistance, increasing problems with the introduction of the consecutive study system in German higher education led to the conclusion that further training was required if universities were to deliver professional teaching. Since the student protests in 2009, if not earlier, the public has also become aware of the considerable difficulties with the Bachelor's and Master's degree programmes (Norton 2018; Merkt et al. 2016).

Both the German Council of Science and Humanities and the Standing Conference of Ministers of Education and Cultural Affairs have taken a view on staff training in higher education. The pedagogical qualifications of university teachers have increasingly become the focus of higher education policy in recent years. In its recommendations on improving the quality of higher education, the German Council of Science and Humanities stated that "higher education institutions [...] need not only more, but also better qualified teaching staff" (*translated from German*) in order to achieve goals such as better success rates, shorter study times and an improvement in the level of graduates' competence (Wissenschaftsrat 2008, p. 65). It therefore recommended the expansion of higher education staff training institutions, a mandatory "qualification programme for all teachers" (p. 66) and continuing professional development for university teachers.

The German Federal Government and the federal states have responded to these demands with the broad-based funding programme *Quality Pact for Teaching*, which has given rise to a wide range of development projects and qualification programmes in higher education. The latter offer teaching staff, who usually have no pedagogical qualification at the beginning of their teaching career (Seidel and Hoppert 2011), an opportunity to grow into their teaching role with qualified support. However, the few studies available on what is usually voluntary participation in training in higher education (e.g. Lübeck 2009; Pötschke 2004) suggest that, despite being urged to take up more training, only a few teachers are embracing the expanded qualifications on offer. One or two studies have already been conducted on the reasons for non-participation (e.g. Flender 2004; Pötschke 2004).

Generally, university teachers must also have an initial qualification in order to work with ESD. However, a number of studies have shown that ESD has played hardly any role in staff training in higher education (Mulà 2017; Kapitulinová et al. 2015). In Germany, too, the relevance of ESD has long been ignored by higher education development and staff training programmes. "ESD [...] has so far been integrated only marginally into the training of university lecturers, so that the awareness and pedagogical skills of university lecturers have not yet been trained to a sufficient extent" (*translated from German*) (Etzkorn 2018, p. 5f.).

In order to implement ESD more widely in HEIs, it is necessary to offer training on ESD for university staff (cf. Etzkorn 2018). The German National Action Plan on ESD states: "It is the responsibility of every university teacher to also concern him- or herself with sustainability/ ESD. University leaders are expected to include regular participation in ESD training programmes in appointment agreements, especially for early-career university teachers. Incentives should be created to encourage emerging researchers and the entire teaching staff to participate in appropriate training" (*translated from German*) (Nationale Plattform Bildung für nachhaltige Entwicklung 2017, p. 55).

Development and Assessment of ESD Competences

ESD requires an action-oriented, transformative pedagogy (Lozano and Barreiro-Gen 2021; Rieckmann 2018; UNESCO 2017; see also Chap. 17), characterised by the following pedagogical principles (Rieckmann 2018; UNESCO 2017; Littledyke and Manolas 2011; Künzli David 2007):

- Learner-centredness and accessibility
- A focus on action and reflection
- Transformative and transgressive learning[2]
- Prioritisation of participation
- Discovery learning
- Networked learning
- A focus on vision
- The linking of social, self-referential and method-based learning with subject-related learning

These pedagogical principles constitute general guidelines for the design of ESD-based learning processes. ESD presupposes a new culture of teaching and thus also learning and must draw on a wide variety of different pedagogical approaches and methods. Teaching and learning methods are needed that correspond to the above principles and thus promote the development of competences. Such methods include service-learning projects that combine classroom sessions with work-based learning, usually in community-based settings, future workshops, systems games and fish-bowl discussions (UNESCO 2017, p. 55). These teaching and learning methods enable learners to become (co-) designers of their own learning process and thus have a direct influence on their own competence development. They also enable learners to take action to promote sustainable development.

If teaching is designed to promote sustainability competences, it is important to check whether this goal is actually being achieved. It is therefore necessary to assess students' competence development (see Chap. 21). The overall aim of assessment is to check whether ESD is in fact effective and is achieving the goals that have been set. Against this background, various purposes can be

[2]Transformative learning aims to empower learners to question and change the way they see and think about the world (Rodríguez Aboytes and Barth 2020). The related concept of transgressive learning (Lotz-Sisitka et al. 2015) goes a step further emphasising that learning in ESD should transcend the status quo and prepare learners for disruptive thinking and the co-creation of new knowledge.

identified for individual-level assessment in ESD (UNESCO 2017, p. 57):

- "Gather information and record learners' progress and achievement toward intended learning outcomes
- Communicate progress to learners, identify strengths and areas for growth, and use this information to set learning goals
- Provide feedback about the success of teaching and learning processes to help plan, implement and improve these processes
- In formal education, guide decisions about the learner's grading and academic and occupational choices".

Competences are expressed through action in certain situations (performance). Therefore, they can only be assessed indirectly. On the one hand, this makes consideration of the environment a central challenge; on the other, it is necessary to take into account the fact that competences can be demonstrated in different ways with different support and thus in different environments. The manifestation of competences changes with the context; the process can be described in terms of development corridors rather than rigid, objective characteristics (Hartig et al. 2008; Leutner et al. 2017).

When selecting assessment methods, attention should be paid to ensuring that they are constructively aligned with the intended learning outcomes and the relevant teaching and learning practices. According to Redman et al. (2021), there are eight types of tools or methods that can be used for assessing students' sustainability competences:

- Scaled self-assessment
- Reflective writing
- Scenario/case test
- Focus group/interview
- Performance observation
- Concept mapping
- Conventional test
- Regular course work

Even though scaled self-assessment is used quite frequently, it cannot be assumed that it will deliver valid findings on the actual development of competences (Redman et al. 2021).

It is very important that assessment is not narrowed down to certain (cognitive) aspects, but that the global transformative potential and the complexity of ESD are also considered (García et al. 2017). This can be ensured by using different assessment methods (mixed-methods approach), "Given the variety of learning objectives and competencies ESD entails, a range of methods is likely to be required to assess learning accurately" (UNESCO 2017, p. 57) (see also Chap. 21). The desire to measure at least some aspects of competences often leads to a focus purely on the cognitive facets of competence (knowledge level). In ESD, concentrating on cognitive aspects alone is problematic, since they do not constitute the core of the competences in question. If competences relating to holistic action and transformation are to be cultivated to enable students and teachers to deal with global complexity, then volition, motivation, values, attitudes and programmes of action must be considered alongside the cognitive aspects (Asbrand and Martens 2013; see also Chap. 6).

Teachers should not only focus on assessment *of* learning but also consider assessment *for* learning and assessment *as* learning. They "should use a mix of traditional assessment methods and more reflective and performance-based methods, such as self- and peer assessment, that capture learners' insights on such aspects as personal transformation, deepened understanding of critical inquiry, and engagement and civic agency. Feedback from educators, peer feedback and self-evaluation (e.g. using reflective journals or portfolios) empower learners to monitor their own learning processes and to identify possibilities for improvement" (UNESCO 2017, p. 57).

Staff Training at the University of Vechta

In the summer term of 2018, ESD staff training was delivered for the first time at the University of Vechta. The university was then involved with the RSP project from autumn 2018, and from that point, the RSP framework was used to guide and develop the ESD training. In the summer semester 2019, a second round of ESD training was provided. Each of the training workshops was attended by 10–15 professors, lecturers and student teachers. A total of 30 different participants attended the staff training workshops over the course of the 2 years.

The staff training workshops introduced participants to the concept of ESD and enabled them to develop ESD competences on the basis of the RSP framework. The training programme reflected the principles of ESD and enabled participants to engage actively with the concepts of sustainability and ESD and to experience ESD itself. Against the backdrop of the RSP framework, the training programme was thus designed to promote the development of ESD competences. Workshop topics ranged from an introduction to ESD to ESD pedagogy to specific pedagogical approaches and methods such as service learning, interdisciplinary learning and future studies methods.

In October 2019, we held a focus group (Krueger 2014; Stewart and Shamdasani 2014; Escalada and Heong 2009) with six participants from the staff training workshops, investigating four themes. Firstly, we examined the learning from the ESD training programme and considered what further improvement and development the programme might need. The second set of questions looked at the RSP competence framework and its strengths and weaknesses. A third set was dedicated to the application of ESD methods within teaching, as a result of what had been learned during the training programme. "Staff Training at the University of Vechta" and "Discussion and Conclusion" section explored the options for assessing ESD-related learning outcomes. The focus group was transcribed and data analysis undertaken on the basis of qualitative content analysis using MaxQDA (Kuckartz 2014).

One learning point from the training programme that was highlighted by the focus group participants, was the complexity of the various competences and the limited potential of training sessions of only 3–4 h to contribute to the development of the comprehensive competences described in the RSP framework. "[...] you can't even learn that in 2 h, you really have to have a lot of practical competence, methodological competence and I don't think that's something you can necessarily expect of every teacher, not at this level anyway".

With regard to the usability of the RSP competence framework, one focus group participant expressed the view that all the competences were logical in themselves and did not contradict each other, but that "almost every one of these competences requires a decidedly subject-focused introduction, and not a topic-based approach, but rather one that starts with a problem, or is situated in the living environment". Another of the concerns about the twelve RSP competences that provoked discussion was under-reflected normativity; when talking about values, for example, the following questions arose: "Which values are we talking about?", "What defines values?" and "How do I assess this?" Overall, focus group participants found it difficult to identify whether the RSP framework was a good representation of all the required ESD competences, or whether more competences needed to be added, or whether the competences that were included might be of little relevance. The RSP competences were said to be "free of contradiction" and were all deemed "relatable".

After discussing the strengths and weaknesses of the RSP framework, participants were asked whether methods had been introduced in the ESD workshops that they thought they would be able to deploy in their teaching or that they were already using. Overall, discussion of this question yielded three main insights:

1. Some of the training sessions did not focus directly on a specific pedagogical approach or method.

2. During some workshops, the time had passed so quickly that it was difficult for participants to remember specific methods that they could apply.
3. Where specific methods were mentioned, teachers had either modified these to apply them or had given some thought to how to apply them.

Another main issue that arose during the focus group discussion was the lack of appropriate assessment formats "[...] because the thought occurred to me that it is actually difficult for teachers, of course, because the formats don't really exist, so we don't have to put students through any kind of competence-based examination, and some, but not all, of the competences could perhaps even be tested in written exams, not necessarily through multiple-choice questions, but perhaps through open answer formats [...]".

Discussion and Conclusion

The theoretical and practical approaches adopted by the German higher education system demonstrate the relevance of ESD training for educators within the higher education sector. According to Leal Filho et al. (2020) and Cebrián et al. (2020), staff training in ESD needs to prioritise the discussion of values, promote understanding the complexity of systems and encourage thinking "outside the box", all with a focus on the future. Another critical aspect for ESD staff training programmes is multiple perspectives. Focusing solely on knowledge acquisition means that the other dimensions ("learn to live together, learn to do and learn to be", Sleurs 2008) are left out and the emerging picture is biased and not holistic. Here, however, it is problematic that the ESD competences frameworks themselves are already less focused on "learn to be" aspects (Corres et al. 2020).

The integrative perspective was also considered in the focus group, with participants taking the view that all the proposed competences were relatable and interlinked, and that competences should therefore not be seen as the acquisition of knowledge in separate areas. With regard to the University of Vechta staff training, there was positive feedback overall on the training programme, the knowledge the programme provided about sustainable development, ESD and ESD-related methods and the (integrative) references to different ESD competences.

However, there are some limitations to the training sessions as such and this related research, the first being the low number of participants, which is indicative of the lack of attractiveness of staff training workshops. As is generally the case with staff training in higher education (Pötschke 2004; Lübeck 2009), the ESD staff workshops were attended mainly by professors, lecturers and student teachers with a high personal interest in ESD, demonstrating that ESD is still only of interest to individuals that are already working in this field or showing an interest in it. In general, there is a lack of commitment amongst higher education teachers and no demand of any kind for further training. Thus, the question arises of how to attract teachers who are less familiar with and/or uninterested in ESD to such workshops in order to have a greater impact on higher education.

Secondly, critical reflection is needed on the opportunities that such workshops offer for competence development. To develop the complex competences required for ESD, deeper learning processes are necessary, and these require more time (Rieckmann 2018; Weinert 2001). For this reason, staff training workshops can only provide the initial stimulus for such competence development. The most that can be said is that training workshops such as these can act as a trigger, highlighting certain aspects of the competences and providing guidance as to where and how they can be further developed. However, a deeper and more reflective process is required to develop the competences themselves. Some of the insights from the analysis of the focus group highlight the difficulties with implementing new ESD methodologies at the University of Vechta—due to lack of time, limited flexibility in what is taught, and the general circumstances relating to seminars.

Finally, the question of how competences such as the RSP values, empathy and creativity competences can be assessed remains open (Vare et al. 2019; García et al. 2017). To ensure that the process is constructively aligned, assessment methods must be chosen that do justice to the complex interplay of ESD/sustainability competences (Redman et al. 2021; UNESCO 2017; see also Chap. 21). Assessment should not only be about the formal grading of students' performance; its primary purpose should be to enable learners to reflect on their own learning process (UNESCO 2017). However, as the results of the focus group show, assessing ESD competences is still seen as something rather difficult and complicated. Teachers lack knowledge and competences with regard to assessment and thus it is clear that there is a need for further training in this area as well.

To further support the integration of ESD in higher education, it is important to integrate ESD into all the structures and programme content of HEIs. This will ensure that HEIs generate knowledge that can make an important contribution to sustainable development. A whole institution approach (Rath and Schmitt 2017; UNESCO 2017; Sterling 2004) needs to be taken to ESD at HEIs: a holistic, open, accessible and reflective process that all can participate in. HEIs should approach ESD content, methods and principles from the perspective of theory, concepts and methodologies, but above all critically and reflexively. Students are key stakeholders since they are network initiators and visionaries and can thus mobilise the potential of HEIs to develop a sustainable outlook. The whole institution approach thus encompasses not only the core areas of teaching and research but also the operation of HEIs, enabling them to become authentic places of learning.

References

Asbrand, B. & Martens, M. (2013). Qualitative Kompetenzforschung im Lernbereich Globale Entwicklung. Das Beispiel Perspektivenübernahme. In: Rode, H., Overwien, B. (eds.), *Bildung für nachhaltige Entwicklung. Lebenslanges Lernen, Kompetenz und gesellschaftliche Teilhabe* (pp. 47-67). Ökologie und Erziehungswissenschaft.

Brundiers, K., Barth, M., Cebrián, G., Cohen, M., Diaz, L., Doucette-Remington, S., Dripps, W., Habron, G., Harré, N., Jarchow, M., Losch, K., Michel, J., Mochizuki, Y., Rieckmann, M., Parnell, R., Walker, P. & Zint, M. (2021). Key competencies in sustainability in higher education—toward an agreed-upon reference framework. *Sustainability Science*, 16(1), 13–29. https://doi.org/10.1007/s11625-020-00838-2.

Cebrián, G., Junyent, M. & Mulà, I. (2020). Competencies in Education for Sustainable Development: Emerging Teaching and Research Developments. *Sustainability* 2020, 12(2), 579

Corres, A., Rieckmann, M., Espasa, A. & Ruiz-Mallén, I. (2020). Educator Competences in Sustainability Education: A Systematic Review of Frameworks. *Sustainability*, 12, 9858.

Escalada, M. & Heong, K.L. (2009). F*ocus Group Discussion.* https://www.researchgate.net/publication/242589494_Focus_Group_Discussion_1.

Etzkorn, N. (2018). *Hochschulen als Leuchttürme für Bildung für nachhaltige Entwicklung – Auf dem Transformationspfad die Strahlkraft erhöhen.* Executive Summary. Berlin.

Flender, J. (2004). Optimierung ja – Weiterbildung nein? Zur Motivation von Lehrenden, ihre Lehre zu verbessern. *Das Hochschulwesen*, 52(1), 19–24.

García, M.R, Junyent, M. & Fonolleda, M. (2017). How to assess professional competencies in Education for Sustainability? An approach from a perspective of complexity. *International Journal of Sustainability in Higher Education*, 18(5), 772-797.

Hartig, J., Klieme, E. & Leutner, D. (2008). *Assessment of Competencies in Educational Contexts.* Hogrefe Publishing GmbH: Göttingen.

Kapitulinová, D., Dlouhá, J., Ryan, A., Dlouhý, J., Barton, A., Mader, M. et al. (2015). *Leading practice publication. Professional development of university educators on education for sustainable development in European countries.* Prague: Charles University Environment Center.

Krueger, R.A. (2014). *Focus Groups: A Practical Guide for Applied Research*; Sage: London, UK.

Kuckartz, U. (2014). *Qualitative Text Analysis: A guide to Methods, Practice & Using Software*; Sage: Los Angeles, CA, USA; London, UK; New Delhi, India; Singapore; Washington, DC, USA.

Künzli David, C. (2007). *Zukunft mitgestalten. Bildung für eine nachhaltige Entwicklung – Didaktisches Konzept und Umsetzung in der Grundschule.* Bern: Haupt-Verlag.

Leal Filho, W., Levesque, V.R., Salvia, A.L., Paço, A., Fritzen, B., Frankenberger, F., Damke, L.I., Brandli, L.L., Ávila, L.V., Mifsud, M., et al. (2020). University teaching staff and sustainable development: An assessment of competences. *Sustainability Science*, 1–16.

Leutner D., Fleischer J., Grünkorn J., Klieme E. (2017). Competence Assessment in Education: An Introduction. In: Leutner D., Fleischer J., Grünkorn

J., Klieme E. (eds.), *Competence Assessment in Education. Methodology of Educational Measurement and Assessment.* Springer, Cham. https://doi.org/10.1007/978-3-319-50030-0_1

Littledyke, M., & Manolas, E. (2011). Education for Sustainability Pedagogy: Ideological and Epistemological Barriers and Drivers. In: W. Leal Filho (Ed.), *World trends in education for sustainable development* (pp. 77-104). Springer: Frankfurt am Main.

Lotz-Sisitka, H., Wals, A. E., Kronlid, D., & McGarry, D. (2015). Transformative, transgressive social learning: rethinking higher education pedagogy in times of systemic global dysfunction. *Current Opinion in Environmental Sustainability* 16, 73-80.

Lozano, R. & Barreiro-Gen, M. (Hrsg.). (2021). *Strategies for Sustainability. Developing Sustainability Competences Through Pedagogical Approaches: Experiences from International Case Studies*. Springer International Publishing; Imprint: Springer. https://doi.org/10.1007/978-3-030-64965-4.

Lübeck, D. (2009). *Lehransätze in der Hochschullehre.* Dissertation, Berlin.

Merkt, M., Wetzel, C. & Schaper, N. (2016). *Professionalisierung der Hochschuldidaktik.* Bertelsmann: Bielefeld.

Mulà, I. (2017). Catalysing change in higher education for sustainable development: A review of professional development initiatives for university educators. *International Journal of Sustainability in Higher Education* 18 (5), 798–820.

Nationale Plattform Bildung für nachhaltige Entwicklung (Ed.) (2017). *Nationaler Aktionsplan Bildung für nachhaltige Entwicklung. Der deutsche Beitrag zum UNESCO-Weltaktionsprogramm.* Berlin.

Norton, L. (2018). Why is the professionalization of university teaching important? In: L. Norton, *Action Research in Teaching and Learning. A Practical Guide to Conducting Pedagogical Research in Universities* (pp. 35-53). Routledge: London.

Pötschke, M. (2004). Akzeptanz hochschuldidaktischer Weiterbildung. Ergebnisse einer empirischen Studie an der Universität Bremen. *Das Hochschulwesen*, 52(3), 94–99.

Rath, K. & Schmitt, C.T. (2017). Sustainability at Universities: Degrees of Institutionalization for Sustainability at German Higher Education Institutions—A Categorization Pattern. In W. Leal Filho, L. Brandli, P. Castro, J. Newman (Eds.), *Handbook of Theory and Practice of Sustainable Development in Higher Education* (Vol. 1, pp. 451-470). Springer International Publishing.

Redman, A., Wiek, A. & Barth, M. (2021). Current practice of assessing students' sustainability competencies: a review of tools. *Sustainability Science* 16, 117–135. https://doi.org/10.1007/s11625-020-00855-1.

Rieckmann, M. (2018). Chapter 2 - Learning to transform the world: key competencies in ESD. In: A. Leicht, J. Heiss & W. J. Byun (eds.), *Education on the move. Issues and trends in education for sustainable development* (pp. 39–59). United Nations Educational, Scientific and Cultural Organization.

Rodríguez Aboytes, J.G. & Barth, M. (2020). Transformative learning in the field of sustainability: A systematic literature review (1999–2019). *International Journal of Sustainability in Higher Education*, 21(5), 993–1013.

Seidel, T., & Hoppert, A. (2011). Merkmale von Lehre an der Hochschule. Ergebnisse zur Gestaltung von Hochschulseminaren mittels Videoanalysen. *Unterrichtswissenschaft*, 39(2), 154–172.

Sleurs, W. (2008). *Competencies for ESD (Education for Sustainable Development) Teachers. A Framework to Integrate ESD in the Curriculum of Teacher Training Institutes*; Comenius 2.1 Project; CSCT: Brüssel, Belgium. http://www.unece.org/fileadmin/DAM/env/esd/inf.meeting.docs/EGonInd/8mtg/CSCT%20Handbook_Extract.pdf.

Sterling, S. (2004). Higher education, sustainability, and the role of systemic learning. In: P. B. Corcoran and A. E. J. Wals (eds.), *Higher Education and the Challenge of Sustainability: Problematics, Promise, and Practice* (pp. 49–70). Kluwer Academic Publishers: Dordrecht.

Stewart, D.W. & Shamdasani, P.N. (2014). *Focus Groups: Theory and Practices.* Sage: London, UK.

UNESCO (2017). *Education for Sustainable Development Goals. Learning Objectives.* Paris: UNESCO. http://unesdoc.unesco.org/images/0024/002474/247444e.pdf

Vare, P., Arro, G., de Hamer, A., Del Gobbo, G., de Vries, G., Farioli, F., Kadji-Beltran, C., Kangur, M., Mayer, M., Millican, R., Nijdam, C., Réti, M. & Zachariou, A. (2019). Devising a Competence-Based Training Program for Educators of Sustainable Development: Lessons Learned. *Sustainability*, 11(7), 1890. https://doi.org/10.3390/su11071890

Weinert, F. E. (2001). Concept of Competence: A Conceptual Clarification. In D. S. Rychen & L. H. Salganik (Eds.), *Defining and Selecting Key Competencies* (pp. 45–65). Hogrefe und Huber.

Wissenschaftsrat (2008). *Empfehlungen zur Qualitätsverbesserung von Lehre und Studium.* Köln: WR.

ESD Competences and Teacher Educators' Professional Development: Findings and Implications of the Cyprus Example

16

Aravella Zachariou and Chrysanthi Kadji-Beltran

Abstract

The policy context of Cyprus is highly supportive of education for sustainable development (ESD). Despite this, there is a danger that the professional development of teacher educators can be overlooked. This chapter outlines research conducted among ESD educators in Cyprus and highlights some of the gaps and aspects in professional development that are particularly valued by teacher educators. Key findings include the need for collaborative professional development, building critical communities locally and internationally. The chapter highlights the value of working with ESD competence frameworks but closes with a reminder of the importance of contextualizing any such framework.

Keywords

Education for sustainable development (ESD) · ESD competences · Teacher educators' professional development

A. Zachariou (✉)
Frederick University, Unit of EESD/CPI, Nicosia, Cyprus
e-mail: aravella@cytanet.com.cy

C. Kadji-Beltran
Frederick University, Nicosia, Cyprus

Introduction

Teacher trainers for education for sustainable development (ESD) actively facilitate the learning of teachers and student teachers (European Commission 2013, p. 8) and act as role models in order to help them understand and realize their roles as multipliers for ESD in the school context. They challenge teachers to question their way of teaching sustainable development (SD) issues and expose them to new pedagogies and participatory approaches that set ESD at the core of teaching and learning and strengthen school-development through the whole institution approach (Adomβent 2012, pp. 8–9). According to Deem and Lucas (2007) if teacher trainers were better prepared, resourced, and supported, better teacher student learning outcomes could be attained.

Although teacher education is a priority action area for building the ESD competences of educators and trainers (UNESCO 2020) teacher educators' professional development (PD) is often absent or overlooked.

This gap between teachers' and teacher educators' PD could be related to the fact that teacher educators' role is complex, demanding, and unclear because of the lack of coherent professional identity (Lewis 2017). An additional challenge to educators' PD on ESD is the interdisciplinary nature and the different backgrounds of ESD educators (Shulman 1987). This

P. Vare et al. (eds.), *Competences in Education for Sustainable Development*, Sustainable Development Goals Series, https://doi.org/10.1007/978-3-030-91055-6_16

creates a puzzlement about which PD courses are most appropriate to offer, in which form, content, pedagogies, by whom, and in which learning environments? The priority attributed to ESD by different educational systems (UNESCO 2017), as well as the place of ESD in the curriculum generates a diversity that poses further obstacles (UNECE 2020, p. 14).

Teacher continuous PD on ESD is expected to promote quality education and quality *in* education for the achievement of the SDGs (see Chap. 9). ESD competences can support quality through engaging learners in transformative learning that will lead to ESD 2 (Vare and Scott 2007). As highlighted in section one of this book, although often situated at the core of the teaching and learning process, the use of competences raises a series of questions regarding their effective integration in formal, non-formal, and informal education contexts. These include the ways they can be meaningfully translated in praxis, how user-friendly they are, the pedagogies used, and the learning outcomes pursued. The lack of consensus resulting from national and socio-cultural factors (O'Flaherty and Beal 2018) and the fact that this is a new field have resulted in a diversity of ESD competence models, an intriguing area of exploration for a scientific community that seeks to provide dynamic, inclusive, and open models of learning for ESD implementation.

Based on this rationale our chapter's aim is to present and discuss teacher educators' professional development in ESD competences and how this affects their professional role as ESD educators. We use *A Rounder Sense of Purpose* (RSP)[1] competence model because of its specific characteristics and attributes; i.e., it is flexible, evolving and dynamic, easy to understand and apply (Zachariou et al. 2019). The model is accompanied by specific learning outcomes and indicative activities, it is easily accessible because of its digital form, is more practical than theoretical, can be integrated in formal and non-formal education context and it refers to every "actor" of education (teachers, trainers, counsellors, inspectors, professionals in education) (Vare et al. 2019; Zachariou et al. 2019).

Professional Development and ESD in Cyprus

ESD is the compass for Cyprus Educational reform. For teacher professional development, ESD is an intrinsic part of the wider policy on professional development, acknowledging that it can improve the education delivered and impact the greater effort for improving learning outcomes (MoEC (Ministry of Education and Culture) 2015). It constitutes an explicit part of educational reform and seeks to support schools in their effort to improve through developing their individual plans and policies (MoEC (Ministry of Education and Culture) 2017).

According to Andic (2020), inadequate or non-existent professional development on ESD competences is closely linked with limitations in the corresponding education policies and institutional support (p. 159). However, this is not the case in Cyprus, as professional development constitutes part of the official policy "as a prerequisite for any innovation and reform of the Cyprus Educational System" (MoEC (Ministry of Education and Culture) 2007, p. 20). The existence of the Education for the Environment and Sustainable Development (EESD) Unit with the mission of "promoting an effective implementation of ESD through a holistic ESD policy … embracing non-formal and informal education…" (MoEC (Ministry of Education and Culture)/CPI (Cyprus Pedagogical Institute) 2016, p. 12) provides ESD a solid basis within the educational system and its policies, thus facilitating actions such as professional development.

Optional and compulsory ESD training for teachers of all levels of education, school principals, and other education stakeholders have been on offer since 2010 in the form of centralized in-service-training, seminars and conferences, school-based ESD seminars, mentoring systems and non-formal education (MoEC (Ministry of Education and Culture) 2017). Teachers' PD

[1] https://aroundersenseofpurpose.eu/.

courses have been redesigned and oriented towards an ESD competence-based approach following the reform guidelines and responding to the emerging challenge for re-orienting teachers' education to quality education.

Methodology

Our case study presents a summary of the results of wider research that scrutinized the level of integration of ESD competences in ESD teacher educators' professional learning.

Our sample included all the staff of the EESD unit (16 women and 5 men) responsible for designing and delivering ESD teacher professional development in Cyprus. They are ESD Educators in Ministry-coordinated EE/ESD Centers, ESD school advisors, ESD teacher educators and responsible for developing ESD material for schools. Fourteen participants originated from Primary and seven from Secondary Education (4 biologists, 2 physicists, and 1 home-economics teacher). All had postgraduate studies in areas such as ESD, outdoor education, environmental protection and management and ecology. Their teaching experience with ESD ranged from 1 to 15 years (9 years average). Most of the participants had received training on ESD competences using the RSP model, during 2018–2020 through programs, workshops, and other activities, organized by the EESD Unit, in collaboration with universities and other agencies.

Participants were administered a structured questionnaire and were invited to consider the questions and provide their answers in writing. Initially they were invited to present and discuss their ideas and perceptions on the concept of competence and then more specifically describe the extent to which the ESD competences and the RSP model had enhanced their work as ESD educators and teacher trainers. They were asked to critically reflect on the training they had received on ESD competences, how this supported their work, and what needs could be addressed through further training. Some questions specifically concerned the RSP model for ESD competences.

Findings

The findings can be categorized in line with the questions that were asked to the participants as mentioned above.

Defining Competences

Participants perceived competences as personal characteristics, qualities, attributes, or skills that help a person reach a final goal and perform their work adequately and efficiently. Some participants added that competences are also a constantly evolving and improving body of knowledge, attitudes, values, emotions, motivations, strategies, and skills. They acknowledged that this dynamic nature of combining and applying knowledge and values while engaging at the same time with emotional factors, skills, and behaviors is necessary for achieving a goal or task. Additionally, they commented that competences refer to the characteristics that equip and encourage people to respond to and achieve visions and set goals. Only one participant referred to the competences solely as skills that enable somebody to achieve a goal.

Further elaborating on their answers, teacher educators identified and discussed properties and characteristics of competences and how these are developed. They supported the idea that teaching about competences is not enough, as 'competence' is an abstract concept. In order to acquire competences, a combination of theory and practice is required through experience and constant reflection.

Training Received

Most of the ESD educators had the opportunity to attend organized PD on ESD: training sessions on ESD competences through research projects, workshops focusing on the use of ESD competences in teaching and institutionalized compulsory school-based ESD courses. Few reported receiving information on ESD competences during their postgraduate studies. Indicative examples of the training received on ESD competences are: (a) a 15 h training program, one conference, and one workshop provided by a research program on the induction of teachers in ESD, (b) a 15 h training on "Educator Competences in learning for sustainability" for EE/ESD Centers provided by the EU research project *A Rounder Sense of Purpose*, (c) a 5 h workshop on how to use competences in non-formal education through examples, (d) a 5 h workshop by UNESCO Associate Schools Network on the use of ESD competences for achieving sustainable cities and communities outdoors, (e) 2 days compulsory school-based trainings.

Participation in the development and delivery of some aspects of the training mentioned was also considered to be a form of professional development.

Impact on ESD Educators' Professional Development

Most participants considered that the ESD competences had improved their professional development concerning ESD due to the well-structured, competence-based framework for approaching SD issues and the guidance provided by the model used for setting learning goals and developing activities. They also mentioned that this approach supported their overall teaching and learning practice and raised their confidence in developing lessons that transfer lifelong competences. Working with competences broadened their thinking in terms of how to study SD issues and raised their teaching and learning standards as well as the engagement and action standards that support a deeper level of ESD. Additionally, competences helped them understand the interconnections and interactions between SD issues and gave them a broader understanding of the SDGs.

Professional learning activities such as the peer reviewing and the feedback process followed for the development of competence-based ESD lesson plans and programs in formal and informal education helped participants develop their collective learning culture. They used a variety of educational techniques and helped each other to develop more engaging, creative, attractive and pleasant activities.

Support Needed

Among the ideas educators gave for enhancing the training offered were professional learning focused on developing material, studying examples of good practice, and providing practical opportunities for understanding the ways in which competences are used. There were also calls for ESD professional education to be longer and more frequent allowing enough time for interaction.

ESD educators believe that their professional learning would be more effective if it were channeled through active learning communities for ESD educators through which they could produce and share material, exchange ideas, and promote collaborations. Learning communities could expand to include colleagues from other countries in order to give an insight into the work done abroad. Such an expansion might take the form of a digital community where learners can post their ideas and discuss them with peers.

Another factor that could support ESD educators' work is the use of action research as a model for the ESD professional programs and tools that they develop.

Familiarity with the RSP Model and Suggestions for Improvement

RSP is the ESD competence model officially used by the EESD Unit of the Cyprus Pedagogical Institute for designing and teaching ESD professional development courses (e.g., workshops, training delivered through research programs) as well as lesson plans and educational material on ESD and SDGs for formal and non-formal education. Most ESD teacher educators became familiar with the model through their work and personal study for the development of various ESD learning activities and programs, depending on the sector and the tasks they are assigned with

(ESD teachers' educators, ESD advisors, ESD educators in Environmental Education Centers).

They considered the RSP model to be an additional tool for the ESD educators and the teachers. With this in mind, they suggested that a more detailed description of the examples would make use of the model easier.

Discussion

As seen above, the educational political context in Cyprus is very supportive of ESD, which is a key contribution to ESD implementation. However, this does not mean that teacher educators' PD cannot be improved. The following discussion, based on the above results, suggests possible directions for improvement.

The majority of the teacher educators appear to perceive competences as a complex of knowledge skills and attitudes that enable successful task performance (Spady 1994), including cognitive, functional, ethical, and personal dimensions (Wiek et al. 2011). The complexity and ambiguity of the concept seems to be translated in their professional learning as 'knowing competences', which refers to what Bruner defined as the 'structure of knowledge' (content knowledge), that is the theories, principles, and concepts of a particular discipline (as cited in Shulman 1992). This perception prevents them from using ESD competences effectively in their professional context, indicating the need for combining theory with practice. From this perspective it is important to include the knowledge of *what* (content knowledge of ESD competences) and the knowledge of *how* (the pedagogical approaches, teaching and methods of delivery of ESD competences in praxis) in teacher educators' professional development. The types of ESD professional development commonly used for teacher educators, e.g., international partnerships within European projects, outdoor learning activities and workshops (UNESCO 2017) and especially school-based seminars and mentoring, have many implications on ESD educators' professional teaching and learning. Through these techniques educators acquire experience, develop their self-efficacy, become more effective and motivated to promote ESD through their professional practice (Kadji-Beltran et al. 2014). This collective learning in communities of peers differs from traditional forms of professional development because learning is not linear and transferable from the trainer to the trainee. The educator has a dual role—educator and learner—and is exposed to reflection and interaction in an in situ learning environment. This is highlighted by the participants, as they acknowledge their double role in developing and delivering professional learning programs as both learners and educators.

Such forms of professional development seem to be related to the overall improvement of teachers' teaching and learning practice, raised confidence in developing competence-based lessons, broadened thinking in terms of how to study SD issues and raised teaching and learning standards on ESD. All these elements strengthen their self-efficacy when teaching about and for ESD competences since they "feel a sense of ownership in designing their lessons and learning environments" (Timm and Barth 2020, p. 9). The importance attributed to peer reviewing, and the critical professional interaction this requires, reveals the importance of re-situating PD for ESD Educators on the basis of professional learning communities. Educators appreciate and crave for a framework through which they can work together and critically interrogate their practice in an ongoing, reflective, collaborative, inclusive, learning-oriented and growth-promoting way (Stoll and Seashore-Lewis 2007).

Peer learning, professional learning communities, digital learning communities, and action research are stressed as critical elements for the improvement of their ESD teaching and learning. Digital communities of learning indicate the importance of digital education to strengthen the pedagogical transformation in the service of teachers and learners and at the same time the need for a broader exposure to the ESD experience, through enabled access to the knowledge and experience of other countries. The need for a broader interaction is connected to the systemic and global nature of sustainability issues, through which local experiences and knowledge can be combined with global communication and collaboration (Sipos et al. 2008; Ricard et al. 2020). Additionally, action research can enhance ESD teachers' professional learning and teaching on ESD competence as part of the four-phased cyclical process of inquiry (plan, action, outcome, and reflection) providing the educators with opportunities to reflect upon, improve, and innovate their practice (Tilbury et al. 2005).

Although some forms of professional development in Cyprus are compulsory, the time and programs devoted to ESD educators' professional development are limited to a few hours and do not correspond to specific long-term planning and goals that can ensure a continuity in the PD of ESD educators. Time alone does not necessarily mean effectiveness as effective professional development needs to be well organized, carefully structured, purposefully directed and focused, addressing both content and pedagogy (Garet et al. 2001). The efforts for connecting competences with the knowledge domain and pedagogical approaches of ESD are limited (Lozano et al. 2019) despite the number of ESD competence models and frameworks available.

Without claiming that the RSP model is the most appropriate framework for competence-based professional development on ESD, teachers' responses revealed that it is user-friendly, useful, and practical. Its value lays in the clarity with which the competences and their interconnections are defined, their links to learning outcomes, and the variety of proposed activities. Its organization has cohesion and consistency and is applicable for diverse settings, forms of education, and key recipients (educators, teachers, professional developers, and policy makers). The structure and the speculations behind this model, its introduction in the Cyprus educational context and its acceptance by the ESD educators raise critical questions regarding the framing of ESD competences and their delivery in ESD educators' professional development: What evidence best reflects the achievement of those competences? How can that evidence be gathered in a valid and meaningful way (Guskey 2000) in order to support designing more complete, holistic, and systemic ESD professional development programs for ESD educators? Despite the positive response on the RSP model, it needs time to be 'intellectually digested' and critically integrated. This finding is aligned with the NSDC (National Staff Development Council) (2001) position, arguing that the most effective professional development comes not from the implementation of a particular set of "best practices" but from their careful adaptation to the specific context, content, and processes.

Final Remarks

Although we explored how a specific ESD competence model was adapted and developed in a national context, it is hoped that the outcomes and discussion contribute to the international dialogue on the professional development of ESD teacher educators.

Despite the benefits of a competence-based ESD at all levels and stages of the educational process, a lot of challenges emerge concerning their effective implementation in ESD educators' professional development. The absence of discussion on the professional development of ESD teacher educators is more evident in the light of the ongoing dialogue on ESD competences in initial teacher education. Teacher education on ESD is part of the systemic effort for a successful integration of ESD in educational systems. It needs to be acknowledged as part of the changes that have to take place if relevant policies are to be enacted effectively. The implementation of ESD models for ESD competences needs to be addressed within this framework. Focus should be placed on how any model will be introduced, for which purpose, and in which educational settings (see also Chap. 13). Care must be taken to adapt the model for the ESD education of educators and how this can respond to their learning needs given their context, pedagogical practices and policies as well as the mechanics of the educational system in which they work.

References

Adomβent, A. (2012). ESD Training of Trainers. Accessed 5 December 2020. https://www.esd-expert.net/files/ESD-Expert/pdf/ESD-ToT-international.pdf

Andic, A. (2020). Continuing Professional Development of Teachers in Education for Sustainable Development-case study of the Republic of Croatia. *Teacher Development*, 24(2):143-164.

Deem, R. & Lucas, L. (2007). Research and teaching cultures in two contrasting UK policy contexts: Academic life in Education Departments in five English and Scottish universities, Higher Education, July 2007, Volume 54, Issue 1, pp 115–133.

European Commission (2013). Teaching and Learning International Survey (TALIS). Main Findings from the Survey and Implications for Education and Training Policies in Europe. Accessed 30 November 2020. https://ec.europa.eu/assets/eac/education/library/reports/2014/talis_en.pdf

Garet, Michael S., Andrew C. Porter, Laura Desimone, Beatrice F. Birman, and Kwang Suk Yoon. (2001). What Makes Professional Development Effective? Results from a National Sample of Teachers. *American Educational Research Journal*, 38: 915-945.

Guskey, Thomas R. (2000). *Evaluating Professional Development*. Thousand Oaks, Calif.: Corwin Press.

Kadji-Beltran C., Zachariou A., Liarakou G., Flogaiti E. (2014). Empowering Education for Sustainable Development in Schools through Mentoring. *Professional Development in Education*. 40 (717-739).

Lewis, H. (2017). Professional Development for Teacher Educators: The Missing Link? Wales Journal of Education,18(3):160–170. https://doi.org/10.16922/wje.19.1.10

Lozano, R.; Barreiro-Gen, M.; Lozano, F.J.; Sammalisto, K. (2019). Teaching Sustainability in European Higher Education Institutions: Assessing the Connections between Competences and Pedagogical Approaches. *Sustainability*, *11* (6). https://doi.org/10.3390/su11061602

MoEC (Ministry of Education and Culture). (2007). *Στρατηγικός Σχεδιασμός για την Παιδεία. Η ολική αναθεώρηση του εκπαιδευτικού μας συστήματος* [Strategic Plan for the Education: A holistic reform if our Educational System]. Nicosia.

MoEC (Ministry of Education and Culture). (2015). *Διαμόρφωση Ενιαίας Πολιτικής για την Επαγγελματική Μάθηση των Εκπαιδευτικών Λειτουργών* [Framing a Unified Policy for Teachers Professional Development]. Nicosia: MoEC.

MoEC (Ministry of Education and Culture). (2017). Υλοποίηση της Ενιαίας Πολιτικής για την Επαγγελματική Μάθηση των εκπαιδευτικών κατά το 2017-2018 [Implementing a Unified Policy for Teachers Professional Development during 2017-2018]. Circular: CPI.7.1.10.8.3/4. Accessed 30 November 2020 http://enimerosi.moec.gov.cy/archeia/1/ypp6177a

MoEC (Ministry of Education and Culture)/CPI (Cyprus Pedagogical Institute). (2016). *Πρόταση για Δημιουργία της Μονάδας για την Εκπαίδευση για το Περιβάλλον και την Αειφόρο Ανάπτυξη* [Proposal for Founding a Unit of Education for Environment and Sustainable Development]. CPI: Nicosia.

NSDC (National Staff Development Council). (2001). *NSDC's Standards for Staff Development*. Oxford, Ohio.

O'Flaherty, J., and M. E. Beal. (2018). Core Competencies and High Leverage Practices of the Beginning Teacher: A Synthesis of the Literature. Journal of Education for Teaching. 44 (4): 461–478. https://doi.org/10.1080/02607476.2018.1450826

Ricard M., Zachariou A., Burgos D. (2020). Digital Education, Information and Communication Technology, and Education for Sustainable Development. In: Burgos D. (eds) *Radical Solutions and eLearning. Lecture Notes in Educational Technology. Springer* (pp.27-39), *Singapore.* https://doi.org/10.1007/978-981-15-4952-6_2

Shulman, L.S. (1987). Knowledge and Teaching: Foundations of the New Reform. *Harvard Educational* Review 57: 1-22.

Shulman, L., (1992). Ways of seeing, ways of knowing, ways of teaching, ways of learning about teaching. *Journal of curriculum studies*, 28 (5), 393–396.

Sipos, Y., Battisti, B., &Grimm, K. (2008). Achieving transformative sustainability learning: Engaging head, hands and heart. *International Journal of Sustainability in Higher Education, 9*(1), 68–86. https://doi.org/10.1108/14676370810842193

Spady, W. (1994). *Outcomes Based Education: Critical Issues and Answers*. American Association of School Administration: Arlington, Virginia.

Stoll, L. and Seashore-Lewis, K. (2007). Professional learning communities, divergence, depth and dilemmas. Buckingham: Open University Press.

Tilbury D, Coleman V, Garlick D (2005) *A national review of environmental education and its contribution to sustainability in Australia: school education.* Australian Government Department of Environment and Heritage and Australian Research Institute in Education for Sustainability: Canberra.

Timm, J.M. and Barth, M. (2020). Making Education for Sustainable Development happen in elementary schools: The role of Teachers. Environmental Education Research. https://doi.org/10.1080/13504622.2020.1813256

UNECE. (2020). Learning from each other: achievements, challenges and ways forward: Fourth evaluation report of the UNECE Strategy for Education for Sustainable Development. Assessed 4 December 2020. http://staging2.unece.org.net4all.ch/fileadmin/DAM/env/esd/15th_Meeting_SC/Doc/Informal_documents/Information_paper_No.2_4th_revised_evaluation_report_after_Oct_2020_meeting_revised_5Nov2020.pdf

UNESCO (2017). A review of Education for Sustainable Development and Global Citizenship Education in

Teacher Education. Background paper prepared for the 2017/8 Global Education Monitoring Report *Accountability in education: Meeting our commitments*. UCL Institute of Education: Liverpool.

UNESCO (2020). Education for Sustainable Development: A roamap#ESDfor2030. Paris: France.

Vare, P., & Scott, W. (2007). Learning for a Change: Exploring the relationship between education and sustainable development. *Journal for Education for Sustainable Development*, *1*(2), 191–198.

Vare P, Arro G., de Hamer A., Del Gobbo J., de Vries G., Farioli F., Kadji-Beltran C., Kangur M., Mayer M., Nijdam C., Millican R., Réti M., Zachariou A. (2019). Devising a competence-based training programme for educators of sustainable development: lessons learned. *Sustainable Education and Approaches* 11 (1890) Sustainability 2019, 11, 1890; https://doi.org/10.3390/su11071890.

Wiek, A., Withycombe L., & Redman, C.L. (2011). Key Competencies in Sustainability: A Reference Framework for Academic Program Development. *Sustainability Science 6*(2), 203–218.

Zachariou A., Kadji C., Vare P., Millican R. (2019). Teacher Professional Development and Education on ESD competences. Implications for Practice. Encyclopedia of Sustainability in Higher Education (Springer) (https://doi.org/10.1007/978-3-319-63951-2_180-1)

Part III

Pedagogy and Competences

Connections Between Sustainable Development Competences and Pedagogical Approaches

17

Rodrigo Lozano and Maria Barreiro-Gen

Abstract

This chapter provides a discussion on the connections between sustainable development competences and the pedagogical approaches used to develop them. It first reviews the competences and pedagogical approaches. The chapter highlights that a combination of pedagogical approaches is needed to better develop sustainability competences. The combination of pedagogical approaches has to take into consideration as well as the context and nature of the discipline, programme, and course being taught. The combination should encourage teachers and their students to challenge traditional mono-disciplinary and silo approaches in order to provide solutions to complex, multi-level, and wicked problems affecting the economic, environmental, social, and time dimensions of sustainability.

Keywords

Education for sustainable development · Competence · Pedagogy

Introduction

Much of modern education is based on Newtonian and Cartesian approaches, based on rationality, causality, mechanistic interpretation, silo thinking, and reductionism (Ketola 2009; Lovelock 2007; Nonaka and Takeuchi 2001). Although such approaches have resulted in unparalleled advances in development and industrialisation (Dunphy et al. 2003; Jensen 1993), the over-reliance on rationality, whilst neglecting and ignoring emotions (Henry 2001), has led us to an unsustainable present and a threatened future (Carley and Christie 2000; Haberl et al. 2011; Reid 1995).

A paradigm revolution is needed to break through existing knowledge barriers and current unsustainable mental models and foster metanoia for sustainability (Lozano 2014). New ways of learning are needed, which actively and consciously engage in the use and protection of natural resources and the safeguarding and improvement of societal well-being for this generation and future ones (Barth and Rieckmann 2012; Burke 2000; Rosner 1995). This revolution requires a radical transformation of the competences that

R. Lozano (✉)
Faculty of Engineering and Sustainable Development, University of Gävle, Gävle, Sweden

Organisational Sustainability, Ltd., Cardiff, UK
e-mail: Rodrigo.Lozano@hig.se

M. Barreiro-Gen
Faculty of Engineering and Sustainable Development, University of Gävle, Gävle, Sweden

P. Vare et al. (eds.), *Competences in Education for Sustainable Development*, Sustainable Development Goals Series, https://doi.org/10.1007/978-3-030-91055-6_17

teachers must have (Barth and Rieckmann 2012; Corres et al. 2020; Vare et al. 2019) and the ones that students must develop (Brundiers et al. 2021; Lozano et al. 2017, 2019; Rieckmann 2018).

State of the Art of Developing Sustainable Development Competences

A number of researchers have discussed the necessary SD competences for students and graduates (Barth et al. 2007; Brundiers et al. 2021; Lambrechts et al. 2013; Vare et al. 2019), and how to develop such competences through pedagogical approaches (Hopkinson and James 2010; Lozano and Barreiro-Gen 2021; Yanarella et al. 2000).

During the last 10 years, there has been a growing body of literature addressing and discussing the definition and use of competences for SD. Competences are a way of describing desired educational outcomes (Hager and Beckett 1995; Segalàs et al. 2010; Sturmberg and Hinchy 2010). They include cognitive, functional, ethical, and personal dimensions (Commission of the European Communities 2005) and link complex knowledge, skills, and attitudes (Wiek et al. 2011). Competences are a way of describing intended educational outcomes (Hager and Beckett 1995; Segalàs et al. 2010; Sturmberg and Hinchy 2010).

Lists of competences relating to education for SD and their use have been proposed by several authors in recent years. Wiek et al. (2011) proposed five overall competence groups (Systems thinking; Anticipatory; Normative; Strategic; and Interpersonal competences). Rieckmann (2012) proposed the following twelve competences: Systemic thinking and handling of complexity; Anticipatory thinking; Critical thinking; Acting fairly and ecologically; Cooperation in (heterogeneous) groups; Participation; Empathy and change of perspective; Interdisciplinary work; Communication and use of media; Planning and realising innovative projects; Evaluation; and Ambiguity and frustration tolerance. Lambrechts et al. (2013) proposed six competences: Responsibility; Emotional intelligence; System orientation; Future orientation; Personal involvement; and Ability to take action. Lozano et al. (2017) proposed twelve competences: Systems thinking; Interdisciplinary work; Anticipatory thinking; Justice, responsibility, and ethics; Critical thinking and analysis; Interpersonal relations and collaboration; Empathy and change of perspective; Communication and use of media; Strategic action; Personal involvement; Assessment and evaluation; and Tolerance for ambiguity and uncertainty.

A combination of different pedagogical approaches is needed to develop SD competences (UNESCO 2006), which allows students to benefit from different learning processes (UNESCO 2012). Pedagogy is defined as "the art or science of teaching" (OED 2007). The choice of pedagogical approach depends on the pedagogical and educational goals and the specifics of the situation (regarding students, teachers, or the learning environment) (de Freitas and Oliver 2005). A variation in pedagogical approaches is important to address the diversity of students (e.g. gender or cultural background) (Ceulemans and De Prins 2010; UNESCO 2006, 2012). Nonetheless, alternative pedagogical approaches to traditional lectures have not yet been widely utilised in education to promote sustainability (Juárez-Nájera et al. 2006; Lozano and Barreiro-Gen 2021; Seatter and Ceulemans 2017).

Recently, there have been proposals on how to connect the competences and pedagogical approaches (Cotton and Winter 2010; Lozano et al. 2017; Segalàs et al. 2010; Sipos et al. 2008). For example, Ceulemans and De Prins (2010) presented a range of student-activating methods (e.g. videos, brainstorming, case studies, team work, jigsaw, assignments, problem-oriented education, oral presentations, and project learning). Lambrechts et al. (2013) identified a number of pedagogical approaches to develop SD competences including the Socratic method, group discussion, role play, group or personal diaries, brainstorming, and peer assessment; internships, solving real community problems, outdoor education, bibliographic research, problem analysis, value clarification, case studies, and concept mapping. Cotton and Winter (2010) proposed the following pedagogical approaches: role-plays and simulations; group discussions; stimulus activities (watching a video or looking at photos, poems, or newspaper extracts to initiate reflection or discussion); debates; critical incidents (students are given an example and asked what they would do, what they could do, and what they should do); case studies; reflexive accounts; personal development planning; critical reading and writing; problem-based learning; fieldwork; and modelling good practice. Lozano et al. (2017) proposed twelve pedagogical approaches divided in three groups: (1) Universal: broadly applicable pedagogies that have been used in many disciplines and contexts (case studies, interdisciplinary team teaching, lecturing, mind and concept maps, and project and/or problem-based learning); (2) Community and social justice: pedagogies developed specifically for use in addressing social justice and community-building (community service learning, jigsaw/interlinked teams, participatory action research); and (3) Environmental Education: pedagogies emerging from environmental sciences and environmental education practices (eco-justice and community, place-based environmental education, supply chain/life cycle analysis, and traditional ecological knowledge).

There has been limited efforts combining pedagogical approaches and competences such as: the case-based approaches for sustainability science (Sprain and Timpson 2012); the effectiveness of different pedagogical approaches in engineering courses for improving student awareness of sustainability (Segalàs et al. 2010); the connections between pedagogical approaches, knowledge domains (declarative, procedural, effectiveness, and social knowledge); and four key competences (systems thinking, foresight, collaboration, and change-agent skills) in the context of primary and secondary education (Frisk and Larson 2011). Lozano et al. (2017) proposed a theoretical framework connecting competences to pedagogical approaches, see Fig. 17.1 where a green cell represents a pedagogical approach that usually contributes to the competence, whilst a yellow cell represents a pedagogical approach that is likely to contribute to the competence. The proposed framework connects the course aims to delivery in ESD by highlighting the connections between pedagogical approaches and competences. The framework is aimed at helping educators in creating and updating their courses to provide a more complete, holistic, and systemic sustainability education to future leaders, decision makers, educators, and change agents.

Fig. 17.1 Framework connecting sustainable development pedagogical approaches to competences. The green cells indicate a high likelihood of addressing the competence, the yellow cells indicate that the approach may address it, and the white cells indicate that the approach does not address the competence. Source: Lozano et al. (2017)

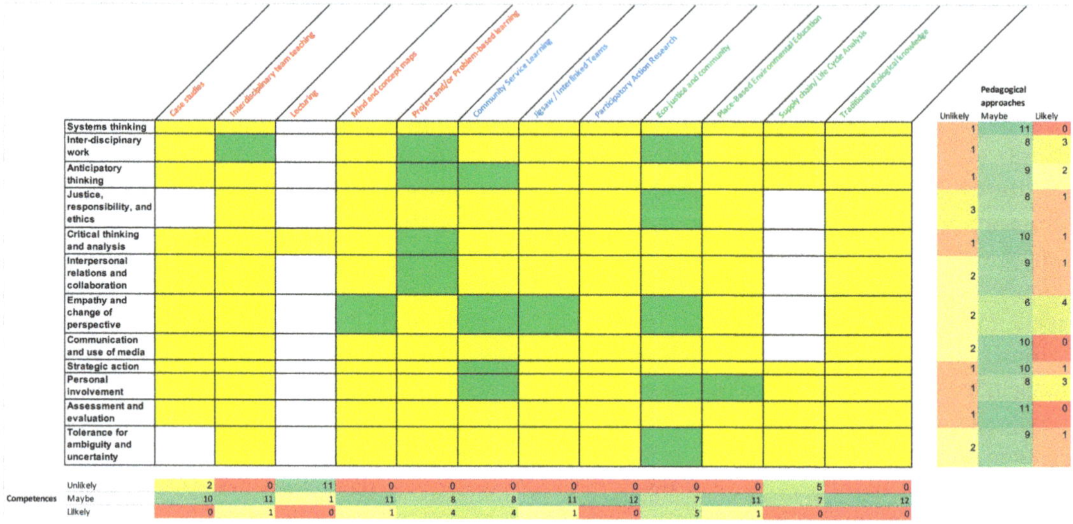

Fig. 17.2 Updated framework connecting sustainable development pedagogical approaches to competences (using Spearman correlation). The green cells indicate a high likelihood of addressing the competence, the yellow cells indicate that the approach may address it, and the white cells indicate that the approach does not address the competence. Source: Lozano et al. (2019)

Lozano et al. (2019) carried out a survey to test the framework connecting sustainable development pedagogical approaches to competences. They received 390 complete responses from teachers at European HEIs (with a response rate of 9.80%). Their results helped to update the theoretical framework, where more pedagogical approaches "may" develop sustainability competences than originally considered. The competences most likely to be developed were, in descending order (Fig. 17.2): empathy and change of perspective; interdisciplinary work; personal involvement; anticipatory thinking; justice, responsibility and ethics; critical thinking and analysis; interpersonal relations and collaboration; strategic action; personal involvement; and tolerance for ambiguity and uncertainty. The pedagogical approaches with the most likelihood to develop sustainability competences were, in descending order: Eco-justice and community; Project and/or problem-based learning; Community service learning; Interdisciplinary team learning; Mind and concept maps; Jigsaw/interlinked teams; and Place-based environmental education. Three pedagogical approaches appear likely to develop the most competences (Eco-justice and community, Project and/or problem-based learning, and Community service learning). This indicates that pedagogical approaches that are more practical and linked to the community appear to have a better likelihood to improve developing sustainability competences.

Discussion and Conclusions

The empirical research on sustainability competences showed that critical thinking and analysis and interdisciplinary work are the ones most widely covered, followed by systems thinking, interpersonal relations and collaboration, and assessment and evaluation (Lozano et al. 2019). These competences tend to be more discussed in the literature (see Rieckmann 2012; Wiek et al. 2011), which may explain their recognition and coverage, whereas the other competences tend to be less discussed in the literature. There is no competence that is considered not important and all should be covered.

The pedagogical approaches most widely used have been lecturing, case studies, and project- or problem-based learning. However, lecturing and case studies were two of the three least effective ways to develop sustainability competences. The other universal pedagogical approaches (project- or problem-based learning, interdisciplinary team teaching, and mind maps) are some of the most effective ways to develop the competences. From the social community and social justice pedagogical approaches the most effective one was community service and learning, followed by jigsaw/interlinked teams. From the environmental education approaches the most effective one was eco-justice and community, followed by place-based environmental education, traditional ecological knowledge, and finally supply chain/life cycle analysis (the one with the least likelihood to develop competences after lecturing). The range of pedagogical approaches was much higher than that of the competences. The pedagogical approaches less used are jigsaw/interlinked teams, place-based environmental education, community service learning, and eco-justice and community; however, they have a good potential to develop the competences, especially eco-justice and community. This shows that although some pedagogical approaches have a good potential to develop sustainability competences, it is necessary to combine them to most effectively cover all the competences.

Although, recently, there has been considerable progress on competences for sustainable development and on pedagogical approaches, there is still work to be done on the connection between how courses are delivered (pedagogical approaches) and how they may affect sustainability competences in order to achieve an educational paradigm revolution, such as authors of this book may wish to achieve.

This chapter presents a framework aimed at providing the bases for connecting sustainability competences and pedagogical approaches. To better embed sustainability into the mindset of our students, it is necessary that the complete set of sustainability competences are developed through a combination of pedagogical approaches.

References

Barth, M., & Rieckmann, M. (2012). Academic staff development as a catalyst for curriculum change towards education for sustainable development: an output perspective. *Journal of Cleaner Production*, *26*(26), 28–36. https://doi.org/10.1016/j.jclepro.2011.12.011

Barth, M., Godemann, J., Rieckmann, M., & Stoltenberg, U. (2007). Developing key competencies for sustainable development in higher education. *International Journal of Sustainability in Higher Education*, *8*(4), 416–430. https://doi.org/10.1108/14676370710823582

Brundiers, K., Barth, M., Cebrián, G., Cohen, M., Diaz, L., Doucette-Remington, S., Dripps, W., Habron, G., Harré, N., Jarchow, M., Losch, K., Michel, J., Mochizuki, Y., Rieckmann, M., Parnell, R., Walker, P., & Zint, M. (2021). Key competencies in sustainability in higher education—toward an agreed-upon reference framework. *Sustainability Science*, *16*(1), 13–29. https://doi.org/10.1007/s11625-020-00838-2

Burke, J. (2000). *The knowledge web* (First). Touchstone.

Carley, M., & Christie, I. (2000). *Managing sustainable development* (Second). Earthscan Publications Ltd. www.earthscan.co.uk

Ceulemans, K., & De Prins, M. (2010). Teacher's manual and method for SD integration in curricula. *Journal of Cleaner Production*, *18*(7), 645–651. https://doi.org/10.1016/j.jclepro.2009.09.014

Commission of the European Communities. (2005). *Towards a European Qualifications Framework for Lifelong Learning*.

Corres, A., Rieckmann, M., Espasa, A., & Ruiz-Mallén, I. (2020). Educator Competences in Sustainability Education: A Systematic Review of Frameworks. *Sustainability*, *12*(23), 9858. https://doi.org/10.3390/su12239858

Cotton, D., & Winter, J. (2010). 'It's not just bits of paper and light bulbs': A review of sustainability pedagogies and their potential for use in higher education. ,. In P. Jones, D. Selby, & Sterling (Eds.), *Sustainability education: Perspectives and practice across higher education*. Earthscan.

de Freitas, S., & Oliver, M. (2005). Does E-learning Policy Drive Change in Higher Education?: A case study relating models of organisational change to e-learning implementation. *Journal of Higher Education Policy and Management*, *27*(1), 81–96. https://doi.org/10.1080/13600800500046255

Dunphy, D., Griffiths, A., Benn, S., Dunphy, D., Griffit, A., Durphy, D., Griffiths, A., Benn, S., Dunphy, D., Griffit, A., Griffiths, A., & Benn, S. (2003). *Organizational change for corporate sustainability* (Third). Routledge.

Frisk, E., & Larson, K. (2011). Educating for sustainability: Competencies & practices for transformative action. *Journal of Sustainability Education*, *2*(March), 1–20. https://doi.org/10.1080/13504622.2012.700697

Haberl, H., Fischer-Kowalski, M., Krausmann, F., Martinez-Alier, J., & Winiwarter, V. (2011). A Socio-metabolic Transition towards Sustainability? Challenges for Another Great Transformation. *Sustainable Development*, *19*, 1–14. https://doi.org/10.1002/sd.410

Hager, P., & Beckett, D. (1995). Philosophical underpinnings of the integrated conception of competence. *Educational Philosophy and Theory*, *27*(1), 1–24. https://doi.org/10.1111/j.1469-5812.1995.tb00209.x

Henry, J. (2001). *Creative management* (Second). Sage Publications Ltd.

Hopkinson, P., & James, P. (2010). Practical pedagogy for embedding ESD in science, technology, engineering and mathematics curricula. *International Journal of Sustainability in Higher Education*, *11*(4), 365–379. https://doi.org/10.1108/14676371011077586

Jensen, M. C. (1993). The Modern Industrial Revolution, Exit, and the Failure of Internal Control Systems. *The Journal of Finance*, *48*(3).

Juárez-Nájera, M., Dieleman, H., & Turpin-Marion, S. (2006). Sustainability in Mexican Higher Education: towards a new academic and professional culture. *Journal of Cleaner Production*, *14*(9–11), 1028–1038. https://doi.org/10.1016/j.jclepro.2005.11.049

Ketola, T. (2009). Pre-morphean paradigm - An alternative to modern and post-modern paradigms of corporate sustainability. *Sustainable Development*, *17*(2), 114–126. https://doi.org/10.1002/sd.406

Lambrechts, W., Mulà, I., Ceulemans, K., Molderez, I., & Gaeremynck, V. (2013). The integration of competences for sustainable development in higher education: an analysis of bachelor programs in management. *Journal of Cleaner Production*, *48*, 65–73. https://doi.org/10.1016/j.jclepro.2011.12.034

Lovelock, J. (2007). *The revenge of Gaia*. Penguin Group.

Lozano, R. (2014). Creativity and Organizational Learning as Means to Foster Sustainability. *Sustainable Development*, *22*(3), n/a-n/a. https://doi.org/10.1002/sd.540

Lozano, R., & Barreiro-Gen, M. (Eds.). (2021). *Developing Sustainability Competences Through Pedagogical Approaches*. Springer International Publishing.

Lozano, R., Merrill, M. M. Y. M., Sammalisto, K., Ceulemans, K., & Lozano, F. J. F. J. F. J. (2017). Connecting Competences and Pedagogical Approaches for Sustainable Development in Higher Education: A Literature Review and Framework Proposal. *Sustainability*, *9*(11), 1889. https://doi.org/10.3390/su9101889

Lozano, R., Barreiro-Gen, M., Lozano, F. J., & Sammalisto, K. (2019). Teaching Sustainability in European Higher Education Institutions : Assessing the Connections between Competences and Pedagogical Approaches. *Sustainability*, *11*(6), 1–17. https://doi.org/10.3390/su11061602

Nonaka, I., & Takeuchi, H. (2001). Organizational knowledge creation. In *Creative Management* (Second). SAGE Publications, Inc.

OED. (2007). *Shorter Oxford English Dictionary* (Sixth). Oxford University Press Inc.

Reid, D. (1995). *Sustainable development. An introductory guide* (First). Earthscan Publications Ltd.

Rieckmann, M. (2012). Future-oriented higher education: Which key competencies should be fostered through university teaching and learning? *Futures*, *44*(2), 127–135. https://doi.org/10.1016/j.futures.2011.09.005

Rieckmann, M. (2018). Rieckman 2018.pdf. In A. Leicht, J. Heiss, & W. J. Byun (Eds.), *Issues and trends in Education for Sustainable Development* (pp. 39–59). UNESCO.

Rosner, W. J. (1995). Mental models for sustainability. *Journal of Cleaner Production*, *3*(1–2), 107–121.

Seatter, C. S., & Ceulemans, K. (2017). *Teaching Sustainability in Higher Education : Pedagogical Styles that Make a Difference*. *47*(2), 47–70.

Segalàs, J., Ferrer-Balas, D., & Mulder, K. F. (2010). What do engineering students learn in sustainability courses? The effect of the pedagogical approach. *Journal of Cleaner Production*, *18*, 275–284. https://doi.org/10.1016/j.jclepro.2009.09.012

Sipos, Y., Battisti, B., & Grimm, K. (2008). Achieving transformative sustainability learning: engaging head, hands and heart. *International Journal of Sustainability in Higher Education*, *9*(1), 68–86. https://doi.org/10.1108/14676370810842193

Sprain, L., & Timpson, W. M. (2012). Pedagogy for Sustainability Science: Case-Based Approaches for Interdisciplinary Instruction. *Environmental Communication: A Journal of Nature and Culture*, *October*, 1–19. https://doi.org/10.1080/17524032.2012.714394

Sturmberg, J. P., & Hinchy, J. (2010). Borderline competence - from a complexity perspective: conceptualization and implementation for certifying examinations. *Journal of Evaluation in Clinical Practice*, *16*(4), 867–872. https://doi.org/10.1111/j.1365-2753.2010.01501.x

UNESCO. (2006). *Framework for the UN DESD International Implementation Scheme – ED/DESD/2006/PI/1*.

UNESCO. (2012). *Education for Sustainable Development Sourcebook. Education for Sustainable Development in Action. Learning & Training Tools N°4 - 2012*.

Vare, P., Arro, G., de Hamer, A., Del Gobbo, G., de Vries, G., Farioli, F., Kadji-Beltran, C., Kangur, M., Mayer, M., Millican, R., Nijdam, C., Réti, M., & Zachariou, A. (2019). Devising a Competence-Based Training Program for Educators of Sustainable Development: Lessons Learned. *Sustainability*, *11*(7), 1890. https://doi.org/10.3390/su11071890

Wiek, A., Withycombe, L., & Redman, C. L. (2011). Key competencies in sustainability: a reference framework for academic program development. *Sustain Sci*, *6*, 203–218. https://doi.org/10.1007/s11625-011-0132-6

Yanarella, E. J., Levine, R. S., & Dumreicher, H. (2000). The space of flows , the rules of play, and sustainable urban design: The sustainability game as a tool of critical pedagogy in higher education. *International Journal of Sustainability in Higher Education*, *1*(1), 48–66.

Art and Connectedness within Sustainability: Educating Through Aesthetic Pedagogies

18

María Heras

Abstract

Different educational approaches have emphasised the relevance of experiential learning when approaching complex and highly dynamic systems and the need to combine different ways of learning, knowing and valuing reality when educating for sustainability. The arts, understood as accessible cultural practices and expressions, can help educators and learners in this journey by offering different lenses to understand and sense our world. Art experiences open up intuitive and non-verbal forms of engagement, drawing on tacit knowledge and emotions as a key source of insight into the dynamics of complex systems. The learning that unfolds from these experiences can be conceptualised as 'aesthetic learning': a kind of experiential learning that is visceral, emotional and intuitive and permits ambiguity, incompleteness, contradiction and complexity, providing a means to express these without reducing them. This chapter explores the potentials of 'aesthetic learning' in sustainability education, and specifically in reinforcing educators' work on sustainability competencies with learners. For this purpose, a diversity of arts-based educational experiences are reviewed in order to explore the potentials and tensions of these emerging aesthetic and critical pedagogies in the work of different sustainability competence frameworks.

Keywords

Aesthetic learning · Arts-based education · Experiential learning · Sustainability education · Competence frameworks

Introduction

> A culture populated by a people whose imagination is impoverished has a static future. In such a culture, there will be little change because there will be little sense of possibility.
> Eisner (2002)

While the Competence Turn (Chap. 2) in sustainability education arrived to bridge knowledge with action in the context of 'wicked' sustainability challenges, it is also acknowledged that to transform individual capacities into real sustainability actions, more than knowledge and skills is needed: values, motivations and opportunities also lie at the core of transformational action (Rieckmann 2018). In this regard, the Competence Turn also involves a rethinking of pedagogies guiding sustainability education, emphasising

M. Heras (✉)
Internet Interdisciplinary Institute (IN3), Universitat Oberta de Catalunya (UOC), Barcelona, Spain
e-mail: mheras0@uoc.edu

P. Vare et al. (eds.), *Competences in Education for Sustainable Development*, Sustainable Development Goals Series, https://doi.org/10.1007/978-3-030-91055-6_18

learner-centred and action-oriented approaches (ibid). In these approaches, underpinned by constructivist theories of learning, learners take an active role in knowledge development through situated and reflexive learning processes that move "from transmissive towards transformative learning" (Sterling 2003, p. 11). Sterling considers *transformative learning* to be "a quality of learning that is deeply engaging, and touches and changes deep levels of values and belief through a process of realisation and recognition" that "inevitably gives rise to a heightened relational sensibility and a sense of ethical responsibility" (Sterling 2010, p. 514). Understood this way, the notion of transformative learning goes beyond cognitive, individual dimensions to include as well relational, normative and affective domains of learning. In this approach, the educator becomes a facilitator who empowers and challenges critically reflective learners to change their worldviews (Rieckmann 2018; Sterling 2010).

Transformative pedagogies emphasise the relevance of experiential learning when approaching complex and highly dynamic systems and the need to combine different ways of learning, knowing and valuing reality when educating for sustainability (Dieleman and Huisingh 2006; Sipos et al. 2008). Feeling and sensing, and not only understanding sustainability as an abstract and distant concept, become crucial in meaning-making and in engaging learners in sustainability transformations (Jickling 2017). Obviously, implementing these pedagogical approaches requires teaching methods and educational experiences aligned with their principles.

In this chapter, I explore the potentials of the arts and 'aesthetic learning' in sustainability education and specifically in reinforcing educators' work on sustainability competences with learners. For that purpose, I review a diversity of arts-based educational experiences and explore the potentials and tensions of these emerging aesthetic and critical pedagogies in the work of different sustainability competence frameworks. Specifically, the following explorative questions guide my inquiry: *How are the arts applied in these experiences and with which motivations? What kinds of insights are reported and how can they critically inform competence frameworks in sustainability education?*

Before that, the following section briefly contextualises the application of the arts within sustainability and environmental education and introduces the notion of aesthetical learning.

Slow, Embodied, Aesthetic Pedagogies: The Artistic Turn in Sustainability Education

The arts, understood as accessible cultural practices and expressions, can help educators and learners in sustainability explorations by offering different lenses to understand and sense our world. Art experiences open up intuitive and non-verbal forms of engagement, drawing on tacit knowledge and emotions as a key source of insight into the dynamics of complex systems (Eisner 2002; Greenwood 2011). Arts' combination of cognitive, embodied, intuitive and emotional awareness and its appeal to open our senses are especially relevant in sustainability education, as they can illuminate the qualitative complexity of sustainability issues (Lehtonen et al. 2020) as well as foster different approaches to learning in highly explorative and motivating ways (Heras et al. 2016). The aesthetic experience resulting from the creation of art or the interaction with artistic practices and artworks, is mediated by the qualities of the arts—emergent, evocative, provocative, expressive, and thus, potentially conducive to new meanings and perceptions (Mantere 2004). At their best, the arts can provoke, unsettle norms and challenge assumptions, while they can also inspire and open up new perspectives through imagination, new connections and reflective thought (Saratsi et al. 2019). Furthermore, the arts can help strengthen emotional bonds between places and people, which lie at the base of personal motives for caring and acting (Inwood 2008).

The blending of the arts with sustainability has not only been encouraged by sustainability practitioners. In the 1960s, the environmental/ecological art movement sought new perspectives and spaces for creative innovation in the face of pressing environmental and political concerns (Saratsi et al. 2019; Gabrys and Yusoff 2012), which also inspired art educators to stress methods specific to art in environmental education. Streaming from such a movement, the term arts-based environmental education was coined in the 1990s to refer to "a form of learning that aims to develop environmental understanding and caring by encouraging participants to become more receptive to sense perceptions and observations through artistic practice" (Van Boeckel 2013, p. 215). Such an approach does not only confirm the interdisciplinary nature of sustainability education but also its processual condition, "instead of framing it as a static content or a goal to be achieved" (Schröder 2018, p. 131).

Whether coming from one side or the other, these approaches share the unfolding of *sensuous, embodied, critical pedagogies* that emphasise sensory experience as a way of relating self to environments and others (Finley 2011) and promote disturbance and disruptiveness. In this way, rather than certainty, they aim at raising critical questions and experiences that help revisit the world in new directions (Eisner 2002). The learning that unfolds from these experiences can be conceptualised as 'aesthetic learning': a kind of experiential learning that is visceral, emotional and intuitive, it "permits ambiguity, incompleteness, contradiction and complexity and provides a means to express them without reducing them" (Greenwood 2011, p. 51). By appealing to our senses and to a refinement of our *organs of perception* (Van Boeckel 2013), these pedagogical approaches call for the slowing down of the educational process "in order to perceive the unknown, the sometimes wild and unexpected" (Mantere 2004, p. 2).

Methodological Approach

For the purposes of my exploration, I reviewed 13 educational interventions worldwide, strategically and systematically selected from a review in the academic search engine Scopus using relevant keywords, to cover the crossing of sustainability and environmental education with art-based approaches.[1] The resulting sample was screened according to several inclusion criteria and the educational experiences selected to include a diversity of sustainability themes (e.g. climate change, connectedness to nature, biodiversity conservation, sustainability challenges), artistic practices (audio-visual, plastic, literary and performing arts), learning contexts (informal and formal), educational phases (from primary to higher education) and target groups (children, youth, adults) (see Table 18.2 in the next section). Rather than expecting to be representative of the universe of current and recent implementations, this exploration aims at illustrating the breadth and depth of aesthetic learning approaches and their different education potentials and challenges in the work of sustainability competencies in education. Further, only educational experiences that included—and were transparent about—evaluation methods were reviewed in order to ensure a reflexive analysis and properly supported insights.

Data from the 13 selected experiences were collected and organised according to: (1) information characterising the educational intervention (e.g. goals, context, number of participants, topics approached or artistic practices engaged with, Table 18.2); (2) information about the evaluation methodology (e.g. data collection methods and analysis strategies); (3) reported outcomes and/or insights of the experiences (Table 18.3). I then conducted a qualitative content analysis to: (1) identify motivations underlying the reviewed educational experiences; (2) link learning and educational potentials of these experiences with competencies for sustainability. The latter analysis was informed by a set of eight predefined themes corresponding to key sustainability competences previously identified in the literature by Rieckmann (2018); see Table 18.1. This framework was selected as it represents a compendium of six competency frameworks developed by

[1] 'Sustainability education' OR 'environmental education' OR 'education for sustainability' OR 'sustainability learning' AND ('arts' OR 'artistic' OR 'arts-based' OR 'embodiment' OR 'artful').

Table 18.1 Analysis categories used to link learning potentials of arts-based sustainability education approaches with sustainability competencies. Compendium of key sustainability competencies proposed by Rieckmann (2018), p. 43–45

Analysis category (sustainability competency)	Definition: the ability to...
Systems thinking competency	...recognize and understand relationships, to analyse complex systems, to perceive the ways in which systems are embedded within different domains and different scales, and to deal with uncertainty
Critical thinking competency	...question norms, practices and opinions; reflect on own one's values, perceptions and actions; and take a position in the sustainability discourse
Anticipatory competency	...understand and evaluate multiple futures—possible, probable and desirable—and to create one's own visions for the future, to apply the precautionary principle, to assess the consequences of actions, and to deal with risks and changes
Normative competency	...understand and reflect on the norms and values that underlie one's actions and to negotiate sustainability values, principles, goals and targets, in a context of conflicts of interests and trade-offs, uncertain knowledge and contradictions
Strategic competency	...collectively develop and implement innovative actions that further sustainability at the local level and further afield
Collaboration competency	...learn from others; understand and respect the needs, perspectives and actions of others (empathy); understand, relate to and be sensitive to others (empathic leadership), deal with conflicts in a group; and facilitate collaborative and participatory problem-solving
Self-awareness competency	...reflect on one's own role in the local community and (global) society, continually evaluate and further motivate one's actions, and deal with one's feelings and desires

(continued)

Table 18.1 (continued)

Analysis category (sustainability competency)	Definition: the ability to...
Integrated problem-solving competency	...apply different problem-solving frameworks to complex sustainability problems and develop viable, inclusive and equitable solution that promote sustainable development—integrating the above-mentioned competencies (overarching competency)

researchers and experts in the fields of sustainability science and education, gathering key sustainability competencies broadly acknowledged in the field as being of particular importance. Here, competencies are understood as individual dispositions to self-organisation which include an interplay of knowledge, capacities and skills, motives and affective elements in interaction (Rieckmann 2012, p. 129).

Results and Discussion

In order to explore the purposes and potentials behind these educational experiences, I first inquired about the motivations to implement them and their specific developments. In a second set, I analysed what outcomes and challenges were reported and how they might connect to sustainability competencies. The next subsections present these two analytical streams and then discuss how these insights can critically inform sustainability competence frameworks.

How Are the Arts Applied in These Experiences and with Which Motivations?

Most of the reviewed experiences apply forms of participatory art, in which participants are actively involved in the creation process. Only

Table 18.2 Summary of the 13 experiences reviewed

Source (reference)	Sustainability topic/s approached	Artistic domain and approach	Intervention (description)	Participants
1. Lehtonen et al. (2020)	Climate change	Applied drama as a process tool	A two-day course to offer first-hand, personal experiences of using drama in sustainable education. It consisted on three drama participatory workshops: process drama work on the global, social and individual aspects of climate change; outdoor drama practice on relations to nature; and reflections through drama practice or work	Drama education teachers + researchers
2. Raatikainen et al. (2020)	Management of meadows and wood- pastures within traditional rural landscapes	Various: creative writing, paintings, movement, crafts (socially engaged art)	One-week school intervention aimed at supporting environmental sensitivity through creative arts-based practices in natural settings, and providing the pupils with a basic understanding of the conservational importance of landscape management. It included 4 workshops (mostly outdoors) and one final public exhibition	5th grade pupils from the rural land + teachers + 1 farmer
3. Bentz (2020)	Climate change	Visual arts (drawings, aquarelle paintings and collages, digitalised)	Art for change project' in a high-school. A experiment with change that invited students to choose one sustainable behavior and adopt it for 30 days. During that time, each student developed an art project about their experience with change. The process also included a transformative learning program that encouraged regular reflection and group discussions	Secondary school students of design communication (art school)
4. Sanchez et al. (2020)	Aquatic ecosystems	Muralism (community art)	Interdisciplinary and participatory environmental education research program involving the painting of several public mural displays as a tool for environmental education. Prior to this, there was a thematic workshop with researchers and environmental educators to discuss about the issues and then one artistic workshop to create the mural together with an artist	Local fishermen and inhabitants + Tachina women + school students

(continued)

Table 18.2 (continued)

Source (reference)	Sustainability topic/s approached	Artistic domain and approach	Intervention (description)	Participants
5. Trott (2019)	Climate change	Photovoice (a participatory action research method using digital photography)	15-week after-school program combining hands-on climate change educational workshops with photovoice to simultaneously explore and expand children's role as agents of sustainable change within their families and communities. After each workshop, children took photographs conveying their perspectives of climate change, which were later translated into sustainability action plans in the wider community	Primary and middle school students from three different locations (separated groups): town, suburbs, city
6. Gray and Birrell (2015)	Landscapes, land restoration	Various (poetic writing, dance, drawings, film, music, artistic notebook)	A year-long place-based enrichment programme using multi-modal creative methods with young participants. The programme connected students with artists and scientists working on an ongoing project of land restoration. Students' dialogue with diverse actors sought to broaden their creativity and provide immersion in the place as a stimulus for artistic creation	Gifted/talented students from primary and secondary school
7. Inwood and Kennedy (2020)	Various (biodiversity; Indigenous knowledge; children environmental rights)	Art installations (as learning medium)	Environmental art education programme over 5 years in a higher education setting with pre-service teachers focusing on how environmental art education can be used to support learning about sustainability. The programme consisted on the creation of site-specific art installations exploring different sustainability topics, to be exhibited in the public space of the school	Pre-service teachers
8. Van Boeckel (2013)	Connection with nature	Clay sculpture (as learning medium)	Different interventions in different countries over several years. An arts-based environmental education workshop in which participants make a small clay sculpture of one's self, a "little-me" with their eyes closed, followed by a group reflective dialogue	Diverse, but all high education adults (some teachers or teacher students)

(continued)

Table 18.2 (continued)

Source (reference)	Sustainability topic/s approached	Artistic domain and approach	Intervention (description)	Participants
9. Heras and Tàbara (2015)	Conservation, natural resource community management	Applied drama and storytelling (as a collective research tool)	One week participatory process to explore young people's perceptions of community forest management and their willingness and barriers to participate. Comprised by a theatrical workshop with young people, as an arts-based research method combining theatrical techniques with storytelling, drawings and guided discussions and group reflections, and the production of an interactive theatrical play, which expanded dialogue to the rest of the community	Teenagers from the indigenous community of Cherán
10. Haynes and Tanner (2015)	Climate change adaptation and disaster risk reduction	Participatory video	A multi-stage process including several workshops where participants were trained in climate change, disaster risk reduction, and film-making; created their own films in the community, identifying priority issues for investigation, and led participatory screening workshops with communities and government officials	Youth from affected communities
11. Manu et al. (2020)	Biodiversity conservation	Textile installations (art as boundary object)	Two textile installation projects inspired by the philosophical concepts and design characteristics of the cubism art movement, representing some of the anthropogenic activities that deplete the biodiversity resources in Ghana. The final installation textile projects are intended to be used by the forest reserves and wildlife sanctuaries for biodiversity conservation education	Faculty members (lecturers and students) from the college of art and built environment + natural park officers
12. Savva et al. (2004)	Different notions of environment (natural, built, cultural)	Artworks made out of materials from the environment	A three-day teachers' in-service training programme, based on the use of the environment as an educational resource. It involved fieldwork applying first-hand experience of the environment (natural settings, rural-building settings, culture and tradition); theoretical lectures about environmental and visual arts education; and critical discussions with participants	Teacher students from nursery, primary and secondary school, with a particular interest in visual arts education

(continued)

Table 18.2 (continued)

Source (reference)	Sustainability topic/s approached	Artistic domain and approach	Intervention (description)	Participants
13. Missiou and Stefos (2012)	Sustainability problems related to deforestation, climate change, chemical hazards and others	Comics, supported by online apps	School project that implied reading and discussing comics of environmental content, and then conducting a small research on the issues approached in the comics. Students also analysed the use of comics to communicate environmental issues and were asked to design and publish their own environmental comics, which were later shared through Facebook	Secondary school students

one experience (textile installations in Ghana) does not involve this hands-on approach and is based on participants' interaction with an artwork already created. Even though many experiences include an exhibition of their resulting artistic outcomes, in most of them the focus is on the art creation process itself (and the possibilities it opens for discovery, perception and expression), rather than in creating a final product of certain aesthetic characteristics. Borrowing the words from Raatikainen et al. (2020), these approaches emphasise art as a dialogic process, being the experiences generated at the core of the artistic-educational results.

Among the different motivations expressed to apply the arts or artistic processes within the experiences reviewed, there are three main themes that can be distinguished (although often interconnected): (1) *increasing knowledge or awareness* of sustainability challenges and socio-ecological dynamics, with an emphasis on stimulating learners' *system's thinking* and awareness of *interconnectedness*; (2) amplifying *personal and affective connections* with the environment, aiming at enhancing learners' *perception and sensitivity*; and (3) promoting *a sense of agency and collective action* through empowering learning experiences, involving, many times, group articulation.

The *relational dimension* is therefore key in these approaches, both understood in analytical and practical terms (i.e. integrated both in contents and pedagogical approaches) and approached cognitively, affectively and sensuously. The arts-based learning experiences reviewed commonly involve inquiry-based pedagogies, opening-up processes of exploration and inquiry, including in some cases participatory action-research, as well as place-based pedagogies, with an emphasis on the spaces where the learning takes place. In this regard, outdoors learning is a recurrent element (8 out of 13 experiences).

What Kinds of Insights Are Reported and How Are They Linked to Competencies?

Table 18.3 shows reported outcomes of these experiences. If we look at these outcomes and insights through the lenses of sustainability competence frameworks, *self-awareness* stands out as the most approached competency, being present in all the reviewed experiences. Learners' self-awareness is approached twofold. First it refers to the *awareness of one's own body* and its feelings and sensations, through the refinement of sensory perception and attentiveness to interdependencies with/within the environment/s and others. Second, awareness is seen as a connection with one's own motivations to act and care, awareness of *being part of* (a community, a society, a common earth), through sense of belonging

Table 18.3 Reported outcomes of the reviewed experiences. Blue letters indicate outcomes reported from educational experiences with teachers or pre-service teachers, while (*) refer to outcomes with both types of participants

Motivations and purposes	Some reported outcomes
Increasing knowledge and awareness of sustainability challenges and socioecological dynamics	Increased awareness of interconnectedness
	Increased awareness of embeded nature connectedness
	Increased perception of the social-ecological complexities of climate change
	Discovery of new perspectives and relationships
	Increased knowledge and awareness of ecosystems' conservation
	Increased awareness of global climate change risks and harmful local impacts
	Changes in cognitive frames
	Integrating a diversity of views and connecting them to individual emotions and motives
	Stimulating critical thinking*
	Eliciting (intrinsic) values of nature
	(Lehtonen et al. 2020, Raatikainen et al. 2020, Bentz 2020, Sánchez et al. 2020, Manu et al. 2020, Trott 2019, Haynes and Tanner 2015, Heras and Tàbara 2015)
Amplifying personal and affective connections with the environment	Raising critical awareness about our ways of being in the world
	Developing individual-level connection to nature
	Acquisition of environmental values and building of sustainable visions of it
	Enhanced sense of place and sense of belonging
	Deeper appreciation of the natural environment and stronger affect towards nature
	Deeper engagement in the site
	Closer connectedness of students with the natural world
	Encouraging a sense of stewardship over the land
	Engagement with one's inner world
	Enhanced awareness of the body and its connections with feelings/sensations
	Expanding personal relationships with environments
	Conveying their own meanings in relation to the environment
	(Raatikainen et al. 2020, Bentz 2020, Sánchez et al. 2020, Gray and Birrell 2015, van Boeckel 2015, Saava et al. 2004)
Promoting sense of agency and collective action Empowering learners	Creating hope, responsibility and care, as well as healing
	Developing a sense of trust* among participants and group cohesion
	Reinforcing sense of belonging to the group/community
	Recognising diversity in the group
	Promoting empathic dialogues*
	Increased awareness as a society, place in the world
	Increased self-confidence*
	Stronger personal creativity
	Stronger beliefs of being "agents of change"
	Overall positive attitude to climate change
	Supporting children's agency
	Expressing learners' own environmental speech, their own voice
	Greater confidence to question community members and decision-makers
	Active involvement in local action
	Increased motivation to pursue action
	(Inwood and Kennedy 2020, Bentz 2020, Manu et al. 2020, Sánchez et al. 2020, Raatikainen et al. 2020, Trott 2019, Gray and Birrell 2015, Haynes and Tanner 2015, Heras and Tàbara 2015, Misious and Stefos 2012, Saava et al. 2004)

and responsibility, through affect. That is, *awareness of individual and group agency* and their role to play in sustainability transformations. These dimensions are closely related to *interpersonal or collaborative competencies* (*n* = 9). While most of the experiences reviewed involved group work or interpersonal dialogues, these exchanges were mostly based on the recognition of different perspectives, resonance with other's experiences and empathy. As connectedness and embedment are important elements in many of the approaches, so are the competencies of *systems' thinking* (*n* = 6), *critical thinking* (*n* = 7) and *normative competency* (*n* = 6). Further, in those experiences based on project development, *strategic and anticipatory competences* (*n* = 4) were approached through the creation of conditions needed to work together (e.g. trust), the recognition of values behind action, the unveiling of tensions and contradictions of different paradigms (development, knowledge, etc.) and ways of being in the world and the assessment of consequences and impacts of different actions.

Finally, five experiences addressed teachers and/or pre-service teachers. Outcomes reported mostly relate to the enhancement of teachers' sensitivity to and personal relationships with the environment and an intensified embodied sense of place. One experience also focused on stimulating teachers' agency, helping them reflect on the roles they can play in sustainability transformations.

As shown in Table 18.3 and discussed in the next section, many of these outcomes become meaningful for the training of sustainability competences when linked with sustainability learning and educational motivations and purposes.

How Can Reported Insights Critically Inform Current Frameworks of Sustainability Competencies?

First, the reviewed experiences reflect the potential of the arts to refine *our sensuous awareness and world engagement*. While sustainability education has placed much focus on sustainability problems (Lehtonen et al. 2020), sensing and connecting can be a first step to love and care, which are seen as preconditions for earth stewardship (Gray and Birrell 2015). In this sense, there is a precious opportunity in arts-based learning experiences to slow down and pay attention to ways of being that are often neglected in educational approaches: perception through our senses, awareness of our inner world and deeper engagement with the environments around us. *Sensing the world* emerges, thus, as a competency that can be developed through the arts, for instance, by reinforcing skills of *attentive listening* (Østergaard 2019), *sensorial imagination* (Van Boeckel 2013) or a 'sensibility to patterns that connect at multiple levels' (Kagan 2011). These are skills that are not just intuitive and they can be unfolded and refined through artistic experiences. Further, there is promising potential in the integration of arts-based approaches within outdoors learning, aimed at establishing personal connections between learners and environments. Although an aesthetic experience can emerge from interactions with the environment without the presence of the arts, arts-based approaches can boost this potential, while placing emphasis in this dimension and offering possibilities to experience and perceive sites differently.

Engaging with the world through the lenses of connectedness and embedment might afford as well to bring a fresh regard to *systems' and critical thinking competencies.* While bringing together cognitive and affective aspects and connecting them to broader socioecological contexts (Raatikainen et al. 2020), arts-based practices can approach complexity and interactions not only as analytical objects of study 'out there' but also as webs of relationships in which we are all involved personally, politically and practically. The questioning of frames and values, perspective taking and creative imagination are some artistic features approaching and confronting systems' complexity, by opening the door to new perspectives and relationships. Indeed, art's capacity to provoke a sense of *estrangement* or moments of de-familiarisation (Van Boeckel 2013) can bring a unique spark to critical and creative thinking in participants, opening-up their learning experiences to emerging meanings, questions and understandings, instead of acting on 'auto-pilot' (ibid).

In approaching sustainability challenges holistically, these experiences also illustrate the potential to work on learners' skills and positive attitudes related to their *agency*, both individually and collectively. Several of the experiences reviewed reported learners' increased awareness of sustainability challenges while also a feeling of empowerment to address them and look at the future with hope. Expressing and processing feelings and emotions can help learners cope with feelings of anxiety or hopelessness, while connecting with nature's beauty and love can connect them with motivations to act and care (instead of being driven by fear).

Further, as observed in the experiences, group work through the arts can help build trust among learners, a sense of group and cohesion and empathic listening skills that are key in any process of collective articulation. In this regard, arts-based experiences in sustainability education can help educators work on learners' *strategic competencies* by opening-up unconventional, fertile spaces for training transversal skills, capacities and attitudes relevant for engaging in collecting action (e.g. communication, listening, respect for others). In such processes, the arts can help educators bring *criticality* and *voices of dissent* (Finley 2011) that are relevant in the development of normative, strategic and collaborative competencies, through the *recognition of diversity and power relations* involved both in sustainability challenges and transformations. Power is a crucial dimension in sustainability commonly neglected in educational approaches, and the arts can contribute to unveil, analyse and disentangle power relationships, both conceptually (while approaching sustainability issues) and practically (within the group that is engaged in the learning process).

Challenges and Needs: What Do Educators Need in Order to Engage with Arts-Based Sustainability Education?

While listing potentials of arts-based practices, there is a latent risk of instrumentalism. Are we tempted to fall precisely into the same prescriptive approach we want to avoid? In designing arts-based educational interventions, are we compromising the inherent value of the arts, which lays on its open, disruptive and emergent nature? As Østergaard (2019) points out, there can be a missed potential in framing arts 'as a tool' instead of a form of knowledge in itself. While pragmatic and certainly efficient to achieve certain goals, an instrumental use of the arts will inevitably miss some of its very intrinsic potentials. Therefore, it is first important to acknowledge that different understandings of the arts and specific implementations and contexts (including who participates) will lead to different experiences, potentials and limitations. The arts are generally recognised as facilitating access and being inclusive but this is not inherent to a designed arts-based activity, and ignoring learners' socio-cultural contexts and power relations can also lead to elitist, manipulative or even alienating learning experiences. In this regard, the potential lays out not only in the integration of the arts, but in the extent to which these practices are contextualised within a transformative learning approach, that is, questioning our beliefs and values, paradigm shifts, connecting practical, political and personal dimensions (Bentz 2020).

Second, arts-based learning experiences require skilled educators able to sustain them. Depending on the personal and professional background of the educator, some artistic practices will be more accessible than others. Many of the experiences reviewed involved drama or arts teachers interested in incorporating sustainability issues in their class. However, there were also experiences involving nursery, primary or secondary school teachers teaching other subjects but with an inclination and a sensibility to be involved in arts-based sustainability education. Obviously, it will be difficult to facilitate a process of deep sensory engagement if the educator has not experienced or developed such sensibility herself. Educators should nurture their own self- and sensuous awareness, listening skills and systemic thinking, as well as bearing an open attitude and curiosity. In this regard, as suggested by the examples in this review, arts-based practices could help educators work on their 'learning to be' competencies, according to the UNECE (2012) framework, such as sense of belonging, attentiveness and emotions management. These competencies have received less attention in the development of pedagogical strategies to promote them (Corres et al. 2020). Further, educators "should be able to bear witness to and hold the space for whatever enfolds in this encounter with artistic process' and 'walk the tightrope between control and non-interfering" (Van Boeckel 2013, p. 8). As a facilitator of learning, the educator engages through arts-based approaches in the dialectic tension of '*active non-activeness*', ensuring that the process can be developed in optimal conditions for learners to fully engage with it, but stepping back once learners are on track (ibid). Although sometimes not easy, this is in line with sustainability education approaches that claim a non-utilitarian or non-prescriptive view of competence-based learning (Sterling 2010). In this way, what might be seen as a challenge, might represent an opportunity as well for educators to refine their own sensorial perception and sensitivity to environments and relationships, and be able to engage in open-ended, emergent educational processes with learners.

Final Thoughts

We live in a world made of relationships (Bateson 1972). At the root of current unsustainability challenges we find unbalanced relationships between different elements of socio-ecosystems and the neglect of the diverse relations and interdependencies that connect us, humans, with all living and nonliving beings in the Earth. Even the disconnection with our bodies—as the first and closest environment we live in, is an example of such neglect. Arts-based sustainability education might offer a chance to reconnect with, re-assess and sense all of these webs of relationships. Through the experiences reviewed, relational and dialogic aesthetics emerge as catalysts of highly engaging experiences where educators and learners can sense their ways of *being in* the world and promote an awareness of interconnectedness, while exploring their capacity to be, to change, to care. Although not prescriptive, these experiences suggest a transformative learning potential that can be untapped by sustainability educators willing to engage in the mystery and open-ended nature of aesthetic experience.

References

Bateson, G. (1972). *Steps to an ecology of mind*. New York: Ballantine.

Bentz, J. (2020). Learning about climate change in, with and through art. *Climatic Change*, 162(3), 1595-1612.

Corres, A., Rieckmann, M., Espasa, A., & Ruiz-Mallén, I. (2020). Educator Competences in Sustainability Education: A Systematic Review of Frameworks. *Sustainability*, 12(23), 9858.

Dieleman, H., and D. Huisingh. (2006). Games by which to learn and teach about sustainable development: exploring the relevance of games and experiential learning for sustainability. *Journal of Cleaner Production* 14(9-11):837-847. https://doi.org/10.1016/j.jclepro.2005.11.031

Eisner EW. 2002. *The Arts and the creation of Mind*, Yale University Press, New Haven and London.

Finley, S. (2011). Ecoaesthetics: Green arts at the intersection of education and social transformation. *Cultural Studies -Critical Methodologies*, *11*(3), 306-313

Gabrys, J. & Yusoff, K. (2012). Arts, Sciences and Climate Change: Practices and Politics at the Threshold. *Science as Culture*, Vol. 21, 1, pp. 1–24. https://doi.org/10.1080/09505431.2010.550139.

Gray, T. & Birrell, C. (2015). 'Touched by the Earth': a place-based outdoor learning programme incorporating the Arts. *Journal of Adventure Education and Outdoor Learning*, https://doi.org/10.1080/14729679.2015.1035293

Greenwood, J. (2011). Aesthetic learning, and learning through the aesthetic. Pages 47-52 *in* S. Schonmann, editor. *Key concepts in theatre/drama education*. Sense Publishers, Rotterdam, Netherlands.

Haynes, K., & Tanner, T. M. (2015). Empowering young people and strengthening resilience: Youth-centred participatory video as a tool for climate change adaptation anddisaster risk reduction.Children's Geographies,13(3), 357–371.

Heras, M. & Tàbara, J.D. (2015): Conservation Theatre: Mirroring Experiences and Performing Stories in Community Management of Natural Resources. Society & Natural Resources

Heras, M., J. D. Tabara, and A. Meza. (2016) Performing biospheric futures with younger generations: a case in the MAB Reserve of La Sepultura, Mexico. Ecology and Society 21(2):14. http://dx.doi.org/10.5751/ES-08317-210214

Inwood, H. J. (2008). At a crossroads: situating place-based art education. *Canadian Journal of Environmental Education,* 13 (1):29-41.

Inwood, H., & Kennedy, A. (2020). Conceptualising art education as environmental activism in preservice teacher education. International Journal of Art and Design Education, 39(3), 585-599.

Jickling, B. (2017). Education Revisited: Creating Educational Experiences that are held, felt, and disruptive. In: (15-30) B. Jickling and S. Sterling (Eds.) *Post-Sustainability and Environmental Education.* Palgrave Studies in Education and the Environment. Cham, Switzerland: Springer Nature.

Kagan, S. (2011). Aesthetics of sustainability: a transdisciplinary sensibility for transformative practices. *Transdisciplinary Journal of Engineering & Science,* 2

Lehtonen, A., Österlind, E., & Viirret, T.L.(2020). Drama in education for sustainability: Becoming connected through embodiment. *International Journal of Education & the Arts,* 21(19). https://doi.org/10.26209/ijea21n19

Mantere, M.H. (2004). *Coming Back to the Senses An Artistic Approach to Environmental Education*: http://www.naturearteducation.org

Manu, G. O., Howard, E. K., Adom, D., & Agyemang, O. (2020). Textile installations inspired by cubism: For biodiversity sustainability education. Journal of Urban Culture Research, 20, 54-68.

Missiou, M., & Stefos, E. (2012). Environmental education through comics and internet applications. A case study in high school. Journal of Environmental Protection and Ecology, 13(3 A)

Østergaard, E. (2019). Music and sustainability education – a contradiction? *Acta Didactica Norge,* 13(2), 2-20.

Raatikainen, K. J., Juhola, K., Huhmarniemi, M., & Peña-Lagos, H. (2020). "Face the cow": reconnecting to nature and increasing capacities for pro-environmental agency, *Ecosystems and People,* 16:1, 273-289, https://doi.org/10.1080/26395916.2020.1817151

Rieckmann, M. (2012). Future-oriented higher education: Which key competencies should be fostered through university teaching and learning? *Futures,* 44, 127–135. [CrossRef]

Rieckmann, M. (2018). Learning to transform the world: Key competences in Education for Sustainable Development. In *Issues and Trends in Education for Sustainable Development*; Leicht, A., Heiss, J., Byun, W.J., Eds.; UNESCO: Paris, France.

Sanchez, E., Vinueza, R., Izurieta, X., & Rey, N. (2020). Use of muralism to promote awareness about aquatic ecosystems and wise water consumption in northwestern Ecuador. Ocean and Coastal Management, 190

Schröder A. (2018) Arts-Based Approaches for Environmental Awareness in University Campuses. In: Leal Filho W., Frankenberger F., Iglesias P., Mülfarth R. (eds) Towards Green Campus Operations. World Sustainability Series. Springer, Cham. https://doi.org/10.1007/978-3-319-76885-4_8

Saratsi, E., Acott, T., Allinson, E., Edwards, D., Fremantle, C., & Fish, R. (2019). Valuing arts and arts research. Valuing nature paper, 22. UK: *Valuing Nature* [online]. Available from: https://valuing-nature.net/valuing-arts-and-arts-research

Savva, A., Trimis, E., & Zachariou, A. (2004). Exploring the links between visual arts and environmental education: Experiences of teachers participating in an in-service training programme. International Journal of Art and Design Education, 23(3), 246-255.

Sipos, Y., B. Battisti, and K. Grimm. (2008). Achieving transformative sustainability learning: engaging head, hands and heart. *International Journal of Sustainability in Higher Education* 9(1):68-86. https://doi.org/10.1108/14676370810842193

Sterling, S. (2003). *Whole systems thinking as a basis for paradigm change in education: exploration in the context of sustainability.* PhD Thesis. University of Bath, Bath, UK.

Sterling, S. (2010). Learning for resilience, or the resilient learner? Towards a necessary reconciliation in a paradigm of sustainable education. *Environmental Education Research,* 16(5-6), 511-528

Trott C.D. (2019) Reshaping our world: Collaborating with children for community-based climate change action. Action Research, 17(1), 42-62

UNECE (2012). *Learning for the Future: Competences in Education for Sustainable Development*; United Nations Economic Commission for Europe: Geneva, Switzerland.

van Boeckel, J. (2013). *At the Heart of Art and Earth: An Exploration of Practices in Arts-Based Environmental Education.* Doctoral Dissertation, Aalto University, Finland, available at: https://aaltodoc.aalto.fi/handle/123456789/13918

19 Gamification for Sustainability: A Possible Renaissance of Role-Playing Game Mechanics in Pedagogy

Mónika Réti, Edit Lippai, and Márk Nemes

Abstract

Education for sustainable development both targets and serves the next generation. Beyond merely offering topics, educators may succeed in developing sustainability related competences in their learners by adapting methods which speak to new generations socialised in the media-landscape of the twenty-first century. Once the confusion around the terms of gamification is clarified, much can be learned from games, especially from role-playing games, in terms of structuring gamified exercises and gaming-for-learning sessions. This short study references the latest results in neuroscience supporting the use of gamification as well as explaining the difference between *educational gaming* and *games for learning*. The aim is not for educators to become authorities ruling over the classroom but coordinators in a playful methodology: partners and participants of shared games and memorable stories written together by all involved.

Keywords

Educational games · Gamification · Neuroscience · RPGs · Sustainability

M. Réti (✉) · E. Lippai · M. Nemes
Hungarian Research Teachers Association, Budapest, Hungary

Introduction

Education for sustainable development (ESD) both targets and serves the next generation. Beyond offering content knowledge, educators may adopt methods that can speak to new generations socialised in the media-landscape of the twenty-first century. Games and gamification are vividly present in students' lives; they shape attitudes and convey models of thought to them. Our core questions are: if we understand gamification as a tool for individualised (formal) learning, to what extent might ESD educators profit from the adaptation or the creation of gamified environments? And assuming efficient ways of gamifying ESD are possible, what sustainability (and ESD) competences would this develop?

Gamification goes back to the most ancient form of learning, that of collaborative, communal storytelling games. During these, the individual (within, with or without a gaming persona) practises skills, builds knowledge and actively solves problems, and all this is permeated by the excitement of getting to know and discover the world—even a (semi-) fictional one, which all the more mirrors the truths of our own reality. This is not meant to be either a self-serving game, or a form of phantasmagorical entertainment, but a controlled progression where the participants are aware that their behaviours within the sto-

P. Vare et al. (eds.), *Competences in Education for Sustainable Development*, Sustainable Development Goals Series, https://doi.org/10.1007/978-3-030-91055-6_19

ryscape, or gaming frame, are lived as experiences. These experiences can be abstracted into scores and descriptors as they further visualise and thus gain control over their shared reality and (if any) gaming personas. While playing and learning, participants may follow their own progress and, in the ideal scenario, are able to choose their own approaches and next steps of their learning path on their own. Of course, educational story-systems are neither fully open, nor completely unguided. It is best to approach such a learning platform with corridor-like 'storytracks' in mind for both educator and learner/player. As long as the participant is made to feel competent and somewhat 'free' to choose from certain authentic options that befit them (and their gaming persona if there is one) in planning their own course of progress, the agency granted will spur them to move forward, *inter alia* through the planned topics.

There have been great misunderstandings about the phenomenon of gamification from the beginning but the world of games itself, and especially video games, is evolving so explosively that definitions constantly lag behind the broadest understanding of common terms and trends. The authors of this article do not consider the use of either toys or board games in the classroom or a point or reward system measuring class performance to be gamification in itself. We contend that gamification occurs when the participants experience the learning process *not* as learning but as a game, an adventure, a process in which they can make meaningful individual choices (and live through the consequences). We envision gamification as a form of active learning, which is societally integrated and which relies on and develops a set of particular competences. While these competences could be identified as a distinct cluster or 'gamification competence', our concern here is to examine whether ESD competences include elements that could be linked specifically to gamification and/or role-playing games (as predecessors of gamified learning environments).

Approaches to Gamification

So what can educators learn from role-playing games and how might educators approach gamification? Becker (2021) sums up several research papers in her simple categorisation: *games* are interactive, they have rules and one or more goals, have a quantifiable measure of progress or success, and have a recognisable ending. It would be a stretch to pick one or two descriptors and shoe-horn them into our teaching practices, but *any* complete methodology that stands up to all of these criteria is gamifiable, that is, it (a) has the potential to contain adaptable mechanics as mini-games to benefit from an *Educational Gaming*-style approach, or (b) might serve as a platform for *Game-for-Learning* sessions. The italicised terms follow Becker's definitions, which are:

- *Educational Games*: a pedagogic approach utilising mini-games integrated into lesson plans, and practically means the injection of game-like elements into formal education.
- *Games for Learning*: often thematic and immersive games which enable professional development and informal learning through playing them.
- *Serious games* are a subset of games designed specifically for purposes other than, or in addition to, pure entertainment (Sawyer 2003). These include games with an agenda (portraying social issues and/or containing a strong political theme or message), health games, advertising games, and, of course, Games for Learning.

Gamification therefore is the use of game design elements in non-game contexts (Deterding 2010), in other words, the individual mini-systems of games in general, or of any single game, taken out and used to power another engine (in our case, education). This highlights a critical requirement: that we must first learn and test the mechanics we plan to introduce in the classroom and/or students' gaming groups, i.e. we should play with the games chosen for adaptation ourselves. We see this as the only way to realise Game-Based Pedagogy (GBP) as educators and enable Game-Based Learning (GBL) for all participants.

Jane McGonigal's 2010 TED talk and her book (McGonigal 2011) kick-started a new wave of interest in gamification but ten years later this phenomenon has become an often-misunderstood buzzword. Against a dense backdrop of gamification contents, we are still unsure how to carefully filter, pick and introduce them into educational practices in a systematic manner. Playful methods, such as simple point systems, guarantee the learner a say in the choice of their developmental path. The most popular games in the world consist of freely selectable quests, and the challenges are completed by the users alone or as part of a team. All the actions of the players are 'banked', which means they get points and symbolic rewards for every achievement they make. Children all around the world know the basic concepts of these systems, such as missions (quests) or treasure (loot). We therefore believe that pedagogy could apply these concepts to great effect within various formal settings. However, this approach still does not appear very often in teacher education (Tóth 2017), despite the fact that a growing majority of students (and a new generation of educators too) are socialised in virtual (gaming) environments that apply such logic.

Ethical Questions

Gamification has undoubtedly revolutionised reward systems; perhaps its biggest impact has been in drawing the attention of several areas, including marketing, education, social media and media design, to the fact that people can be kept constantly activated in almost any process through a design that is set to their needs, interests and motivations (Richter et al. 2015). It has been demonstrated (Alshammari 2021) that gamified design can keep audiences engaged through persuasive design strategies including morally ambiguous captologic programming that utilises psychological tricks and instincts to control attention and consumption. This occurs even if the audience is not actually (or at least, initially) interested in the product or the message being sold (Lucero et al. 2006) including in secondary (Cohen 2016) higher educational settings (Rahman et al. 2018).

There is clearly a moral dimension to such endeavours. We might question for example, whether it is fair to keep people in a constantly attentive (and impulsive) state through such game-inspired interaction. Gamification in education may differ from these gamified designs in that it is tailored to fit certain time frames as well as the needs of the individuals and groups (Tóth 2017) but in this context, ethical considerations must come to the fore, especially with regard to gamified ESD. The aim is not to sell something, not to transfer information or to change attitudes, but to provide a landscape for experimenting with new habits, and to enhance the learner's preparedness to acquire new knowledge connected to sustainability.

Neurobiology has revealed complex molecular mechanisms behind game playing (Bateman and Nacke 2010); a set of hormones and other neurotransmitters have been associated with gamification (Nacke et al. 2011). Although the 'dopamine loop' phenomenon linked to digital games is not supported by any scientific evidence (Pacewitz 2015), it is proven that the brain's self-rewarding systems operate during a set of circumstances (McClelland et al. 1987) including playing games (Van Rooij 2011). This mechanism is 'the biochemical explanation of affirmation of life' (Vizi 2006, p. 108) and one of the key molecules of the reward system is dopamine (Koepp et al. 1998). The non-synaptic neurotransmitters, such as dopamine, responsible for mood, attention, level of arousal, general excitability and motivation (Vizi et al. 2010) establish the brain's readiness for forming strong memories during gamification. This explains why gamification can be a powerful tool in shaping attitudes and influencing behaviour although there are genetic differences within the human population in its effect (Richter et al. 2013). Designing gamified learning environments creates a special setting in the brain's operation with strong affective traits in which the student is prone to experiences shaping their attitudes towards actions suggested by the educator. This means that gamification is not a method particularly or exclusively applicable for ESD, but this also suggests that such a method could be efficiently used for transformative education while bearing in mind the ethical concerns raised above. This in turn places a responsibility on the educator to consider seriously how far to exploit the gamified means in order to achieve the desired transformative ends among their learners (a question also raised in Chap. 6).

Game Design: Experiments with Teachers and Students

When designing a series of *educational mini-games,* or *gaming sessions for learning*, it is crucial to differentiate between these two. Imagine the line between full frontal, direct instruction and purely entertaining games as a spectrum. Firstly, there is no ideal midway point. In our experience, any product that promises to strike a perfect balance between the two becomes unfulfilling as either game or education, while promising to be both. Educators need to decide which way they want (and are permitted) to go. If they want to make education more playful, we are speaking of playful or gamified education through educational games (EG); if, on the other hand, they want to play the kind of games with study groups which also possess educational depth, we will be looking for games for learning (G4L). There are several shades in each direction but the choice needs to be made nevertheless because these are two distinct roads with their own rulesets and advice. In terms of ESD, we recommend using G4L otherwise, instead of learning, entertainment is likely to dominate with little learning outcome, this in turn is unlikely to lead to sustainability competence development.

Secondly, if one wishes to adapt a game to teaching, one must choose the proper game-set. From now on, we will focus more on a special, traditional type of game, which can be both picked for gamifying components or used to provide immersive platforms for story-driven learning, that is, role-playing games (hereinafter: RPGs). Computer-RPGs (CRPGs) are generally single-player focused and even if they grant options such as local multiplayer or co-operation between different players, the general themes, directions and interactions are heavily framed and focused by professional designers. Members of the Hungarian Research Teachers Association experimented with various RPGs to create gamified learning environments for ESD (Lippai 2019; Réti and Lippai 2021). Even simple RPG settings functioned well in the classroom, and proved to be efficient in developing ESD competences. Teachers involved in the pilots found that RPG settings made it easier, and 'more natural', as one of them reported, to work with competences including those concerned with futures, empathy, creativity, action, decisiveness and values. In most RPG game-sets making decisions and dealing with possible future scenarios are part of the process. Using these settings for developing ESD competences allows the educator to reflect specifically on these competences.

Thirdly, the process needs to grant agency for players. After defining the purpose and tailoring it to the audience, it should be considered what 'progress' means and how a reward structure could support that. All this contributes to the game design. However, each well-functioning gamification system, in its final iteration, is a product of co-operation between the community, the instructor and the game environment; that is, the setting (the fluff) and the ruleset (the crunch). The result is a system, which although grounded in unanimously accepted rules, is in constant interaction with the participants and their actions.

The Game Master designs (often cutting and pasting gaming supplements) environments, non-player characters, situations, challenges and battles with the educational themes in mind as well as considering the players' needs and their characters' drives—two important aspects that are always to be handled separately. Since this is already a lot of work, it is strongly recommended to use a pre-set setting. In such a setup, learners/players decide both *what* they want to do and *how* they want to do it. Clearly, game mastering challenges the coordinator to be deeply engaged in these multiple dimensions while remaining open to adapt and improvise as they work towards their and the lesson's goals.

Educational topics may suggest the genre options for games: sustainability might imply a young-adult montage of survival in scarcity, diplomatic challenges, resource-management, base-building and social struggles. In the design process of such narratives, players must be granted agency to go against the directions imagined by the coordinator: the game turns sour if turned into a didactic vehicle for the lessons plan. However, if granted freedom of choice and burden of consequence, players invent new ways to solve the situations portrayed, and their learning process will feel personal and natural.

The peak of gamification is turning the mission (the gamified lesson) into a story, which merges points and rewards with relevant, educational framings. Sustainability themes are plentiful in game settings of alternative histories and alien worlds, science-fiction, utopias, dystopias and especially in post-apocalypse settings. ESD educators are not necessarily expected to use these fantastic scenarios for gamified lessons. In our experience short stories in conventional settings or with a 'crew' of ordinary people are just as impactful and keep the storytelling focused and grounded in real characters instead of a vast, unreal world. Each generation of learners has its own aesthetic and its particular set of social narratives and heroes; one can leverage this to portray (and achieve) different socio-pop-cultural effects. A good game coordinator/educator understands the tastes of their audience and strives to build a story that is grounded in their reality, to motivate and authentically engage the community as a whole.

In our Hungarian experiments in pre-service teacher training, gamification proved to be an efficient tool to develop a set of (individually chosen) competences of student teachers. Assuming that student teachers, teachers and students possess some level of ESD competence, then one aim of teaching ESD could be to develop other ESD competences (or competence components). Dynamic competence models (Sleurs 2008) emphasise the context dependency of competences. They suggest that in practice, educators use a complex mix of competences and in particular situations, some competences are advanced over others. Game settings model these situations and gamification provides opportunities for individually tailoring the development of some competences more strongly than others. This 'palette approach' received critiques from educators who claimed that student teachers would opt for tasks that relied on their existing competences. However, in practice, the Game Master with the reward system or the storyline (the challenges) can guarantee that development takes place where needed. For example, new task types receive a multiplier or the game setting implies complex challenges that require specific competences (Réti et al. 2021). Similarly, in schools, teachers used the same method for developing the students' sustainability competences, which had been pre-selected by students at the individual level, without losing focus on where improvement was needed. In one experiment involving participatory action research with 18 teachers from 10 schools in Hungary (Réti and Lippai 2021), teachers offered challenges linked to 12 different ESD competence areas to students. These challenges were embedded in the story flow of a storyline-based module. Students were allowed to opt for challenges individually for themselves and for their groups. While delivering tasks, students worked on competence areas previously identified as 'to be developed' because those tasks were more rewarding. In other cases, careful selection of student groups by the teacher ensured that different competence areas (linked to tasks chosen by group members) complemented each other. Both approaches assured that targeted competence areas were developed.

Conclusion

We believe that differentiation and apt use of gamification tools for particular functions, lectures and student groups will vastly improve the student's learning (and one's teaching) experience with positive outcomes, in terms of competence development, for both students and teachers. Mini-games stimulate the mind and inspire conventional personal efforts to fulfil formal tasks, while roleplay strengthens communal bonds, grants a chance to try out roles and archetypes, and allows participants to venture into, live through, update and 'fix' problems in a way rarely observed in conventional learning environments. We see a world of opportunity opening up in such practices, within or without establishment walls—but it will be up to educators to learn their most appropriate application and inspire their development in favour of teaching, learning, and mostly playing, better together.

References

Alshammari, A. (2021). Captology in game-based education: a theoretical framework for the design of persuasive games. *Interactive Learning Environments.* https://doi.org/10.1080/10494820.2021.1915803 Accessed 4 Jul 2021

Bateman, C., & Nacke, E. L. (2010). The neurobiology of play. *Futureplay '10: Proceedings of the International Academic Conference on the Future of Game Design and Technology,* May 2010 p 1–8 https://doi.org/10.1145/1920778.1920780

Becker, K. (2021). What's the difference between gamification, serious games, educational games, and game-based learning? In: *Academia Letters.* Article 209. https://doi.org/10.20935/AL209. Accessed 20 Jun 2021.

Cohen, S. J. (2016). *Investigating the impact of gamification on student performance in a secondary science classroom.* (A professional paper submitted in partial fulfilment of the requirements for the degree of Master of Science in Science Education), Montana State University, Bozeman, Montana, 2016. https://scholarworks.montana.edu/xmlui/bitstream/handle/1/10039/CohenS0816.pdf?sequence=5. Accessed 5 Aug 2021.

Deterding, S. (2010) *Pawned. Gamification and Its Discontents.* Presentation to 'Playful 2010', London. http://www.slideshare.net/dings/pawned-gamificationand-its-discontents. Accessed 20 Jun 2021.

Koepp, M. J., Gunn, R. N., Lawrence, A. D., Cunningham, V. J., Dagher, A., Jones, T. & Brooks, D. J. (1998). Evidence for striatal dopamine release during a video game. *Nature*, 393(6682), 266–267.

Lippai, E. (ed) (2019). Egymástól tanulunk 2019. A Kutató Tanárok Országos Szövetségének konferenciakötete. HRTA, Budapest https://kuttanar.hu/altalanos/efop-3-4-4/egymastol-tanulunk-kotet-2 Accessed 4 Aug 2021

Lucero, A., Zuloaga, R., Mota, S., & Munoz, F. (2006). Persuasive technologies in education: improving motivation to read and write for children. In: W. IJsstelstejin et al. (eds) *PERSUASIVE 2006*, LNCS 3962, p 142-153, 2006. Springer-Verlag Berlin Heidelberg 2006. http://www.funkydesignspaces.com/publications/2006/lucero06_read.pdf. Accessed 5 Aug 2021

McClelland, D. C., Patel, V., Stier, D. & Brown, D. (1987). The relationship of affiliative arousal to dopamine release. *Motiv Emot* 11, 51–66 (1987). https://doi.org/10.1007/BF00992213. Accessed 20 Jun 2021

McGonigal, J, (2011). *Reality Is Broken: Why Games Make Us Better and How They Can Change the World.* London: Penguin Books

Nacke, E. L., Bateman, C., Manrdik, L. R. (2011). *BrainHex: preliminary results from a neurobiological gamer typology survey.* ICEC'11: Proceedings of the 10th international conference on Entertainment Computing, October 2011 p 288–293 https://doi.org/10.1007/978-3-642-24500-8_31

Pacewitz, K. (2015). The dopamine loop and its discontents. Analysis of "gamification by design" as biopolitical power/knowledge. In: Kopeć J, Pacewicz K (eds) *Gamification. Critical approaches*. University of Warsaw. p 27-39

Rahman, R., Ahmad, S., Hashim, U. R. (2018). The effectiveness of gamification technique for higher education students engagement in polytechnic Muadzam Shah Pahang, Malaysia. *Int J Educ Technol High Educ* 15, 41 (2018). https://doi.org/10.1186/s41239-018-0123-0. Accessed 4 Aug 2021

Réti, M., & Lippai, E, (2021) Tanulási környezetek tervezésének támogatása a fenntarthatóságra nevelésben ('How to support the planning of learning environments while teaching sustainability'). Paper accepted at the Hungarian Psychological Association 29th National Scientific Assembly, Székesfehérvár, 26-28 August 2021

Réti, M., Lippai, E., Nemes, M. (2021) Nevelői kompetenciák játékosított fejlesztése a fenntarthatóságra nevelésben ('Gamified development of educators' competences while teaching sustainability'). Paper accepted at the Hungarian Psychological Association 29th National Scientific Assembly, Székesfehérvár, 26-28 August 2021

Richter, A., Richter, S., Barman, A., Soch, J., Klein, M., Assmann, A., Libeau, C., Behnisch, G., Wüstenberg, T., Seidenbecher, C.I. and Schott. B.H. (2013). Motivational salience and genetic variability of dopamine D2 receptor expression interact in the modulation of interference processing. *Front. Hum. Neurosci.* 7:250. https://doi.org/10.3389/fnhum.2013.00250

Richter, G., Raban, D. R. & Rafaeli, S. (2015). Studying gamification: the effects of rewards and incentives on motivation. In: Reiners, T., Wood, L. (eds): *Gamification in education and business.* 2015, Springer, p 21-46

Sawyer, B. (2003). *Serious games: improving public policy through game-based learning and simulation.* http://www.seriousgames.org/images/seriousarticle.pdf. Accessed 21 Jun 2021

Sleurs, W. (2008). *Competencies for ESD (Education for Sustainable Development) teachers. A framework to integrate ESD in the curriculum of teacher training institutes*. Brussels. https://unece.org/fileadmin/DAM/env/esd/inf.meeting.docs/EGonInd/8mtg/CSCT%20Handbook_Extract.pdf Accessed 20 Jun 2021

Tóth, M. (2017). *A fejlesztő értékelés lehetőségei.* Scientific thesis. Faculty of Psychology & Pedagogy, Eötvös Lóránt University, Budapest

Van Rooij, A. J. (2011). *Online Video Game Addiction. Exploring a new phenomenon* [PhD Thesis]. Erasmus University Rotterdam, Rotterdam, The Netherlands. https://core.ac.uk/download/pdf/55816439.pdf. Accessed 20 Jun 2021

Vizi, E. S. (2006). Kábítószerek - a kreativitás mítosza és a rombolás valósága (Drugs of Abuse - the Myth of Creativity and the Reality of Destruction). In: Hitseker M, Szilagyi Z (eds) *Mindentudás Egyeteme*, vol. 6. Kossuth Kiadó, Budapest. p 108-125

Vizi, E. S., Fekete, A., Karoly, R. & Mike, A. (2010). Non-synaptic receptors and transporters involved in brain functions and targets of drug treatment. In: *British Journal of Pharmacology.* 2010 Jun; 160(4): 785–809. Published online 2010 Feb 8. https://doi.org/10.1111/j.1476-5381.2009.00624.x

Outdoor and Sustainability Education: How to Link and Implement Them in Teacher Education? An Empirical Perspective

20

Nadia Lausselet and Ismaël Zosso

Abstract

This chapter explores the nexus between place, connection, and sustainability and how this can be addressed within teacher education. It starts by identifying (a) a set of characteristics for transformative outdoor education (TOE) that has the potential to contribute to sustainability education (SE) and (b) the related competences that teachers might need. It then shows how these competences are developed in two courses provided by the largest teacher education institution in French-speaking Switzerland. The analysis of empirical data offers an overview of competences that students appear to have developed within these modules and how, according to them, this relates to sustainability education. The discussion concludes with a reflection on features that can be supportive when working on teachers' competences in TOE so that it contributes to SE.

Keywords

Outdoor education · Teacher competences · Transformative sustainability education · Pedagogical setting

N. Lausselet (✉) · I. Zosso
University of Teacher Education State of Vaud, Lausanne, Switzerland
e-mail: nadia.lausselet@hepl.ch

Introduction

The increasingly perceptible impacts of climate change, the COVID pandemic, and other challenges related to the Anthropocene question the human-nature relationship anew and return the sustainability debate to the forefront. Outdoor education (OE) has long been considered a meaningful approach for working on this human-nature relationship within the context of sustainability; indeed, Lozano et al. (see Chap. 17) confirm that place-based environmental education can foster various competences in sustainability education (SE).

However, not any type of OE can be considered meaningful within the context of sustainability. As Hill (2012) mentions, activities such as those based on adventure pursuit and personal development require further development to contribute to sustainability education (SE). This is not new, as "calls for transformative approaches to outdoor education which embrace human/nature relationships, concepts of sustainability and critical perspectives on gender and class issues, have appeared in the literature since the 1990s" (Hill 2012, p. 18). Based on this, Hill and Brown (2014) have explored ways to combine transformative, outdoor, and sustainability education, working on the "nexus between place, connection and sustainability" (p. 229).

This chapter looks specifically at how this nexus can be addressed within teacher education.

P. Vare et al. (eds.), *Competences in Education for Sustainable Development*, Sustainable Development Goals Series, https://doi.org/10.1007/978-3-030-91055-6_20

It therefore reacts to Nicol et al. (2007) observation that OE intentions remain too 'paper-based' and that the quality of training offers in the field is difficult to assess (Nicol et al. 2007). It also addresses Hill and Brown (2014) belief that further investigation is needed regarding the "impacts on student learning, transformation, and actions" when studying this nexus (p. 229). This chapter starts therefore by identifying (a) a set of characteristics for a transformative outdoor education (TOE) that has the potential to contribute to SE and (b) the related competences[1] that teachers need. It then shows how these competences are developed in two courses provided by the University of Teacher Education Vaud (HEPVD), based in French-speaking Switzerland. The analysis of empirical data offers an overview of competences that students appear to have developed within these courses and how, according to them, this relates to SE. The discussion concludes with a reflection on features that can be supportive when working on teachers' competences in TOE so that it contributes to SE.

Transformative Outdoor Education within the Frame of Sustainability: A Conceptual Framework

Contemporary understandings of OE combine emancipatory educational traditions and current ecological issues while often still referring to Priest's (1986) definition that claims OE, as a method:

> …is in the tradition of experiential learning as advocated by Comenius, Rousseau, Pestalozzi or Dewey; is vital to learning; crosses cognitive, affective and motor domains; takes into account that reality is interdisciplinary in nature; and sees learning as the result of many interactions (pp. 13–14).

Within this broad framework, Lausselet and Zosso (forthcoming) have outlined a transformative outdoor education (TOE) within the frame of sustainability, in an attempt to enhance the potential for OE to contribute to SE. This is based on literature and the experience of working as teachers and teacher educators over years, making some elements of Priest's definition more explicit, and adding a political dimension. It thus echoes Gruenewald's (2008) idea of a critical pedagogy of place and Hill and Brown's (2014) work on the need to combine transformative, outdoor and sustainability education in which sustainability is understood as a "socio-ecological approach (…) that necessarily includes the political" (p. 220). Related teacher competences were derived from these characteristics, taking into consideration teachers' competences for a quality OE[2] identified by Bortolotti (2018). This framework leaves aside logistical aspects and focuses on pedagogical components, complementing Bortolotti's work with a focus on 'sensory fieldwork' (Job et al. 1999) and an affective connection to place that "may assist in the development of an ethic of care" (Hill and Brown 2014, p. 228), both tending towards Rosa (2018) idea of 'resonance'. It also underlines the importance of keeping record of the outdoor experience so as to mobilize it again indoors (Adamina 2010). While a more detailed account of this work is described in Lausselet and Zosso (forthcoming), it is summarized in Fig. 20.1.

This approach to OE works on an intimate relationship with place and articulates it with a collective transformative process, which makes it consistent with SE. The role of the teacher here is both central and in the background: central, because although the place is at the heart of the process, the teacher remains essential in creating the pedagogical repetition and in moderating the process; in the background because it requires adopting a non-transmissive posture leaving room for a real encounter between learner and

[1]We understand competencies as defined by Weinert (2001): the ability and motivation to mobilize content knowledge, skills, and attitude in order to solve a problem.

[2]Teachers must: (a) master the curriculum in order to match the outside world to the school's general objectives; (b) master the logistics related to a different teaching environment; (c) be able to relate to actors in the field; (d) be able to adapt to changing contexts inherent in the outside world; (e) be able to use active, experiential and/or project-based teaching methods.

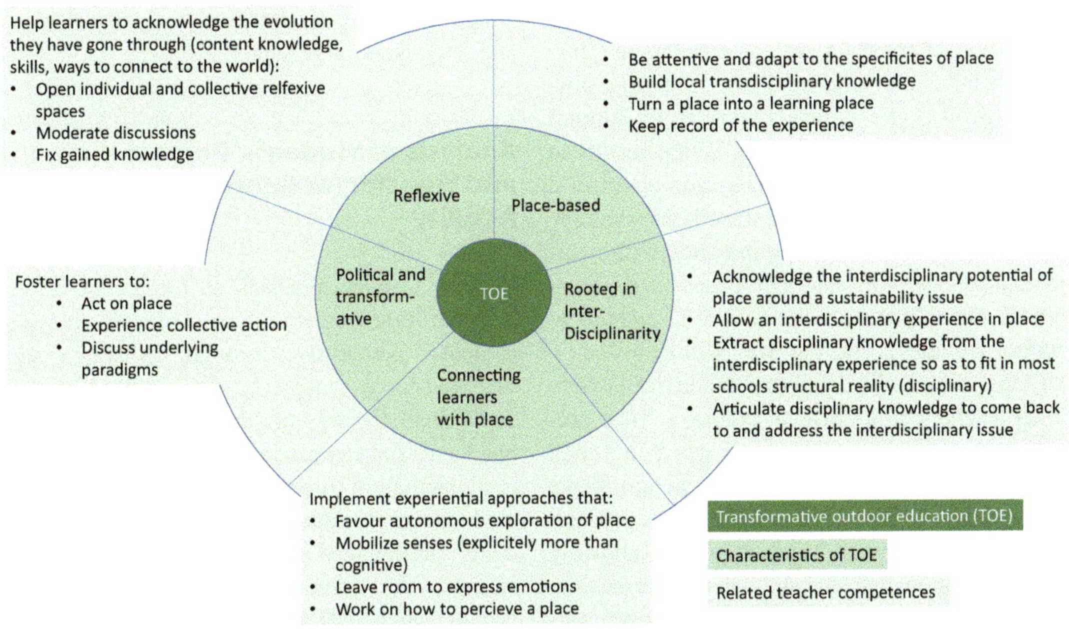

Fig. 20.1 Characteristics of transformative outdoor education and related teacher competences

place. This demands a high level of professionalism which needs to be developed: turning any place into a learning place is not self-evident and must be learned. We now turn to a pedagogical setting that aims to empower student teachers in the operationalization of a TOE and thus in the development of related competences.

Operationalizing Transformative OE: An Outdoor Project-Based Approach

Based on these considerations, we have elaborated a pedagogical setting for two similar outdoor education courses given at HEP Vaud, one for primary education and another for secondary education. It centers on the idea that competences, here related to TOE, cannot be taught but have to be developed by taking action and reflecting on it (Weinert 2001, see also Chap. 15). It thus takes the form of a project-based approach allowing student teachers (from here on 'students') to experience a collective process while elaborating, testing, and disseminating outdoor activities within the conceptual framework presented above. This approach echoes the necessity to empower teachers to be SE project leaders within their schools, capable of working in collective and interdisciplinary dynamics, which implies that they have to be given the opportunity to participate in actual projects and receive reflective feedback during their training (Lange 2011). The aim is thus to develop "individual competences based on a common experience" (*ibid*, p. 74). The role of the teacher educators is to initiate and accompany a process in tension between learning and production, between structure and spaces of freedom, between a place-based experience and transferable knowledge, and between an individual and collective learning process (Lausselet and Zosso 2018). The whole training design seeks, "through exploratory and prospective work" to push "the existing limits," to get out of "routines, traditions, and established customs" in order to contribute to the evolution of today's school (Lange 2017, p. 355).

The courses are part of the interdisciplinary elective pre-service courses offered by the institution. The students are either generalists (primary school) or come from various disciplinary backgrounds (secondary school), most of them being trained in at least two disciplines. Their understanding of outdoor education is mostly reduced to brief fieldtrips having a recreational dimension, or consisting of a visit based on transmissive information delivery (e.g., visit to a sewerage plant). The course's aim is to support students in operationalizing TOE that results in various outdoor activities being combined in a 'pedagogical pathway' (a real trail, for example in the Alps, in a vineyard or in a city, with maps to indicate the way and QR codes at different places along the path that can be scanned to access activities). This pathway will be tested by an actual class, edited and then made available to local teachers on an official platform.[3] The process happens in a safe learning space in which students can experiment, feel, take action, make mistakes, participate in and reflect on the process. The course starts with 2 days outdoors, allowing for exploration, questioning, and experiencing. The rest of the module alternates between outdoors and indoors, group work and collective work, as well as action and reflection. It links on the one hand with the conceptual framework presented earlier and on the other hand with the issues encountered when implementing the approach with pupils. At the end of the semester, explicit links are made with SE. Sustainability models and the Rounder Sense of Purpose (RSP) competence framework (Chap. 5) are explored and related to the students' outdoor activities. For the assessment, students deliver the outdoor activities they have been working on and participate in an oral discussion. For this, they individually imagine a new outdoor activity that they could implement in their class (transfer activity), and reflect on what they learned and how it relates to SE. The transfer activity aims at verifying whether what is declared as learned is actually mobilized in a concrete example. It thus tackles the gap identified by Preston (2016) between the discourse on outdoor education practices that increasingly *pretend* to be learner-centered, mobilizing a sensory and affective dimension, and *actual* practices that remain strongly teacher-led with relatively little autonomy for pupils and which rarely focus consciously on the sensory and affective dimensions.

Analysis of Students' Point of View and Transfer (or Implementation) Activity

In order to have feedback on these courses and possible improvements to be made, we analyzed students' perspective through declarative elements in focus groups (Markova 2003; Gerrisen 2013) held towards the end of semester on the one hand, and examined the transfer activity, i.e., implementation in real classroom settings imagined for the assessment, on the other. All the students participated, implying a group of 18 students for primary (divided into 3 focus groups), and 12 for secondary (divided into 2 focus groups). In the focus groups, a first set of questions examined what students considered as significant learnings and in what ways these would influence their upcoming teaching activity. A second set of questions focused more specifically on the link with SE. In both cases, the questions were left quite open to let whatever seemed meaningful to the students to emerge. The discourses were then interpreted through content analysis (Paillé and Mucchielli 2010) based on our conceptual framework of TOE. In parallel, we evaluated the transfer activities through the double lens of this framework on the one hand and of the coherence with students' declarations in the focus-group discussions on the other. The following trends emerge from the data analysis:

Foster an Active Encounter with Place

All the students considered themselves able to turn a place into a learning place, allowing an active encounter between the learner and a place before making related knowledge explicit; about a quarter of them mentioned that it made them reconsider the role of a teacher. Enough time to immerse themselves in the place, to reflect and to exchange ideas with others, as well as having the possibility to test the activities with learners,

[3] For an example in French, see https://dfjc-files.sos-ch-gva-2.exo.io/s3fs-public/2021-01/SentierDesEquilibres.pdf.

were deemed supportive elements. In the transfer tasks however, around a third of students focused on observation work strongly directed by the teacher, giving the learners a rather passive role that does not allow for a real interaction with place. Moreover, the emotional dimension and the importance to gain a good local knowledge only appeared occasionally in the transfer tasks.

Only a minority of students talked about the possibility of putting learners in a transformative posture by stimulating their imagination and actually experiencing their own impact on a place, and none integrated that transformative dimension in their transfer tasks.

Link Outdoor Experience and School Knowledge

Students regularly mentioned the ability to articulate outdoor and indoor learning, thanks to the record kept of the outdoor experience or the discourse built around it that could then be mobilized in class. The acknowledgment that OE could contribute to the official curricula was considered as a strong motivational factor. This was reflected in the transfer tasks, which even broadened the reflection by including community knowledge through ways not specific to outdoors, e.g., by focusing on parent's knowledge to be worked on in class. The interdisciplinary potential of outdoor work also seemed clear to all but the transfer tasks were mostly rooted in a single discipline. Geography was most often mentioned at primary level, although interdisciplinarity would be easy to implement. For secondary, the link was logically made to the discipline the student was trained in, with possible links to one or two other disciplines. In addition to content knowledge, cross-curricular abilities such as 'cooperation' appeared to be covered by default but no one consciously taught these or made the related learning explicit.

Work with Collective Intelligence

The project-based co-constructed learning process, allowing students to experience a collective contribution to a product that would be useful to others, was seen by all students as a strikingly new perspective on education. Some felt that for the first time they had really learned to collaborate because they needed to contribute to a common and concrete objective by stimulating each other and by using the "power of our collective intelligence" (S1a-14:56). Working outdoors seems to have facilitated this collective dimension by relaxing the pedagogical relationship with the teacher trainers and amongst themselves. The students observed similar dynamics during the test phase with a class. However, quite a few students said that they would not be able to reproduce this kind of iterative collective work, either because they admitted their tendency to be more at ease with transmissive approaches and to want to immediately validate the learners' contribution, or because they did not feel capable of being sufficiently reactive and flexible to facilitate such open learning processes. This is reflected in most transfer tasks, where learners are not given an active, emancipatory role with a collective dimension.

Make an Explicit Link to SE

The vast majority of students were able to make an explicit link to SE, and to evaluate in some ways the contribution, or absence of contribution, of their outdoor activity within SE. They connected to content knowledge around sustainability issues worked on through their activity (e.g., energy, biodiversity), to the idea of fostering a positive and caring bond with the environment or to the fact that OE allowed to realize and measure the impact of human activities on place. Interestingly, future secondary teachers, especially in science, tended to be more focused on content knowledge, whereas future primary teachers and those from secondary dealing with artistic approaches mentioned the caring bond more often. The explicit link to SE competences and to the RSP model was made only occasionally, with mentions of the competences 'systems', 'attentiveness', and 'transdisciplinarity', all of which are part of the competence cluster 'thinking holistically'.[4]

Most students claim to have become conscious of OE as a possibility for tackling SE and have considered this as a motivating factor for implementing the latter. At secondary level, possible ways to integrate SE through OE in language, mathematics, geography, or science were mentioned. At primary level, some students consider OE as an organizing approach allowing the articulation of various disciplines around a common sustainability issue, thus helping to tackle interdisciplinarity. The potential to confront school knowledge to "real-world knowledge" (and vice-versa) was also mentioned, which was considered especially important for SE regarding the complexity of the related issues. Following a similar idea, one student stressed the relevance of interacting with local stakeholders within OE in order to get to know various perspectives and thus tackle complexity. One person even considered that outdoor education, being closer to the children's everyday life, was a powerful means to increase the impact of SE into children's daily routines. Around half of the student teachers mentioned how OE was only one approach, to be combined with indoor approaches in order to contribute to SE.

Against this, some students provided only superficial statements, such as "it helps to sensitize children to sustainability issues" (BPb–6:23) while two students saw the collective action of the project-based approach as the biggest contribution to SE as it empowered learners to act, independently from OE.

In summary, all students claim to have evolved in their attitude and pedagogical know-how, as well as in their motivation, for place-based OE, which indicates that related competences seem to have evolved. They stated a will to combine indoor and outdoor learning, as well as more emancipatory approaches supporting collective learning. Even students returning from an Erasmus semester in the Nordic countries, who have experienced OE regularly, appreciated this structured pedagogical approach that helped to implement OE. Various links to SE were made including the will to implement it through OE, but the political dimension only came up occasionally and was related in part to the project-based approach. Overall, the courses seem to have initiated a deeper reflection on education and the students' role as teacher, at least for some of them, with one stating that:

> I was impressed to see that it is possible to teach differently. I was complaining about the study plan and everything, and now I see that even within this frame, we can do things differently. (S1c-1:12)

Discussion

Firstly, to put these results into perspective, we should note that these were elective courses so the students were per se motivated by OE, thus resolving the question of 'willingness' addressed by Shephard (see Chap. 6). The focus groups also took place before the oral exams, which might imply that students wouldn't dare say negative things, even if this exam didn't count greatly compared to the work done during the semester. With this in mind, the following discussion links this research more specifically to competences in SE.

Prerequisites for a TOE, such as the ability to turn a place into a learning place, can be considered acquired by most students. The gap between the discourse and part of the transfer tasks remains for some students, in line with what Preston (2016) has observed, although conscious efforts were made to overcome it. This demonstrates the cumulative nature of competence acquisition: as such, more attention should be paid to transfer issues, with more exercises to imagine new tasks for various school contexts.

The political transformative aspects, central to TOE and SE, are nearly absent: the idea that it is possible to be a change agent within a place has not really come through despite examples being given. Various hypotheses could explain this:

[4]The RSP competence framework is organized along three competence clusters: thinking holistically, envisioning change, achieving transformation, see Chap. 5 or https://aroundersenseofpurpose.eu/.

- the course hasn't focused enough on this point.
- the belief that teachers are supposed to be 'neutral', as stipulated in the school regulations, plays a hindering role.
- sustainability is not always understood as something political.
- perceiving their pupils as change agents is so far from the students' usual conceptions that one course is not enough to make these conceptions evolve.
- working in a transformative perspective implies being at ease with open and iterative learning processes, which can be perceived as very destabilizing.

In all cases, a more explicit focus might usefully be put on the competences relating to 'envisioning change' and 'achieving transformation'. As these competences are unusual for teachers, a closer attention could also be paid to existing habits and ways to overcome them, as well as on ways to work on students' reactivity in open learning processes.

Nevertheless, even without the political dimension being mentioned, students did multiple links between TOE and SE. This echoes Curnier's (2017) perspective that OE gives the opportunity to work on a multiplicity of knowledges, skills, and attitudes, to tackle the link between knowledge, action, and impact in a concrete way, and to work on the bond between the learner and their environment so as to enhance personal involvement. Each student seems to take something different according to their preconceptions of OE, of SE, or of education more in general, and according to the type of outdoor activity they have developed. This multiplicity could be made more explicit, thus broadening the spectrum of possibilities for everyone. A more careful attention could also be paid on ways to articulate each SE competence specifically with a TOE, as only those relating to "thinking holistically" have been mentioned.

In brief, a TOE seems to be a good base for addressing the potential of OE for SE, but can be reinforced by tackling specific aspects of the latter more in detail on the one hand, and by working on the articulation between both approaches more systematically.

At a more general level, although the courses didn't seem to contribute to a 'transformational education' in itself, the motivation generated and questioning of the teacher's role may yet contribute to the 'transformation of education' that we need (Sterling 2001). A long-term follow-up would be needed to verify this assumption.

Conclusion

We have seen that the assumption that OE contributes to SE competences is easily asserted but not true per se. Even within a type of OE consciously framed within sustainability as presented here, it remains an ambitious task to empower student teachers to implement a TOE contributing to SE. It requires working on a double set of competences, one in TOE, the other in SE. Models addressing SE competences now exist and benefit from wide discussion but they naturally do not cover the specific competences of OE. Models of TOE competences are rarer and less stabilized. There is therefore work to be done on *which* competences contribute to better synergies between OE and SE at the level of teacher education. A conceptual framework such as the one designed here seems a path worth exploring further, as is Hill's model looking at changes needed in values, pedagogical practices, and institutional settings (Hill 2012).

In terms of *how to* develop these competences, more work is needed with students to explicitly articulate TOE and SE, to facilitate the transfer between what has been experienced on the course and what happens in a variety of classroom settings as well as the ability to facilitate open learning processes and to deconstruct existing conceptions of what learning and education means. This means differentiating, for example, between contexts such as those of a secondary science teacher and a primary teacher. Although these elements may seem evident and reflective of general considerations in teacher education, they are "hidden hindrances" not to be forgotten. Moreover, the whole process requires time and requires learning over years, including courses for in-service teachers and post-support experience as stated by Brown (2010). This learning progression for teacher education echoes the learning progression for pupils, implying a "competency double decker"—an underdeveloped idea that is the subject of another article (Lausselet and Zosso forthcoming). Finally, the question of *how to assess* the mentioned competences remains a field that can benefit from further investigation.

References

Adamina, M., (2010). *Ausserschulische Lernorte* (ASLO), [Online]: http://docplayer.org/38514377-Ausserschulische-lernorte-aslo-1.html (accessed on 24.06.2021).

Bortolotti, A. (2018). La ricerca internazionale in outdoor education: una meta-analisi critica. In R. Farné, A. Bortolotti, M. Terrusi. *Outdoor education: prospettive teoriche e buone pratiche*. Roma: Carocci, 61-80.

Brown, M. (2010). Transfer: Outdoor adventure education's Achilles heel? Changing participation as a viable option. *Australian Journal of Outdoor Education, 14*(1), 13–22.

Curnier, D., (2017). *Quel rôle pour l'école dans la transition écologique? Esquisse d'une sociologie politique, environnementale et prospective du curriculum prescrit.* Thèse de doctorat. Lausanne: Université de Lausanne.

Gerrisen, A. (2013). *Focus Group Discussions – a step-by-step guide*. [Online]: https://fr.slideshare.net/AnnetteGerritsen/fgd-manual-14-april-2011 (accessed on 03.07.2021).

Gruenewald, David A. (2008). The best of both worlds: a critical pedagogy of place. *Environmental Education Research*, 14 (3), 308-324.

Hill, A. (2012). Developing approaches to outdoor education that promote sustainability education. *Australian Journal of Outdoor Education*, 16(1), 15-27.

Hill, A., Brown, M. (2014). Intersections between place, sustainability and transformative outdoor experiences. *Journal of Adventure Education & Outdoor Learning*, 14 (3), 217-232.

Job, D., Day, C., & Smyth, T., (1999). *Beyond the bikesheds: Fresh approaches to fieldwork in the school locality*, Sheffield: Geographical Association.

Lange, J-M., (2011). Éducation au développement durable : éléments pour une problématisation de la formation des enseignants, *Carrefours de l'éducation*, 3 (1), 71-85.

Lange, J-M., (2017). Curriculum. In : A. Barthes, J-M. Lange & N. Tutiaux-Guillon (Eds). *Dictionnaire critique des enjeux et concepts des "éducations à"*. Paris: L'Harmattan, 351-359.

Lausselet, N. & Zosso, I., (2018). Projektarbeit an ausserschulischen Lernorten - ein Beitrag zur Bildung für Nachhaltige Entwicklung? In: P. Gautschi, A. Rempfler, B. Sommer & M. Wilhelm (Eds.), *Aneignungspraktiken an ausserschulischen Lernorten: Tagungsband zur 5. Tagung Ausserschulische Lernorte der PH Luzern vom 9. und 10. Juni 2017.* Zürich: LIT Verlag, 173-181.

Lausselet, N. & Zosso, I. (forthcoming). Bonding with the world: A pedagogical approach. In Rolf Jucker & Von Au Jakob (Eds.). *Outdoor-based learning - How can it contribute to high quality learning?* New York: Springer.

Markova, I. (2003). Les focus groups. In S. Moscovici & F. Buschini (Eds.), *Les méthodes des sciences humaines.* Paris : PUF, 221-242.

Nicol, R., Higgins, P., Rossi, H. & Mannion, G. (2007). *Outdoor education in Scotland: a summary of recent research.* [Online]: http://www.docs.hss.ed.ac.uk/education/outdoored/nicol_et_al_oe_scotland_research.pdf (accessed on 09.07.2021).

Paillé, P. & Mucchielli, A. (2010). *L'analyse qualitative en sciences humaines et sociales*. Paris: Armand Colin.

Preston, L., (2016). Field 'Work' Vs 'Feel' Trip: Approaches to Out-of-Class Experiences in Geography Education, *Geographical Education*, Vol. 29, 9-22.

Priest, S., (1986). Redefining outdoor education: a matter of many relationships. *Journal of environmental education.* 17 (3), 13-15.

Rosa, H., 2018. *Résonance: une sociologie de la relation au monde*. Paris : Editions La Découverte.

Sterling, S. (2001). Sustainable education: Re-visioning learning and change. Foxhole, Devon: Green Books

Weinert, F. E. (2001). Concept of Competence: A Conceptual Clarification. In D. S. Rychen & L. H. Salganik (Eds.), *Defining and Selecting Key Competences*. Cambridge: Hogrefe and Huber, 45-65.

Assessing Sustainability Competences: A Discussion on *What* and *How*

21

Francesca Farioli, Michela Mayer, Rick Millican, Aaron Redman, and Paul Vare

Abstract

This chapter comprises a discussion between three international teacher training contexts in Italy, the USA and the UK and explores the tensions that exist in assessing education for sustainable development (ESD) competences. It recognises the need for a form of assessment that is consistent with the aims and values of education for sustainability and that acknowledges the complexity of competences in this field. Debates around the concept of measurement are considered as well as what it is that is being assessed and how judgements are made. Ultimately, the lack of an evidence-based, 'perfect' method is acknowledged along with the need for further research to find an approach that might be applied consistently and reliably in order to provide evidence that those assessed have the competence to make change happen.

Keywords

Education for sustainable development (ESD) · Sustainability · Competence · Competency · Assessment · Measurement

F. Farioli · M. Mayer
IASS-Italian Association for Sustainability Science, Rome, Italy

R. Millican · P. Vare
University of Gloucestershire, Cheltenham, UK

A. Redman (✉)
School of Sustainability, College of Global Futures, Arizona State University, Tempe, AZ, USA
e-mail: aaron.redman@asu.edu

This chapter comprises a discussion between three international teacher training contexts in Italy, the USA and the UK and explores the tensions that exist in assessing education for sustainable development (ESD) competences.

Paul/Rick Colleagues, as you know, all of us have been using educator competences as a way of introducing ESD to educators, trainers and student teachers. Here at the University of Gloucestershire where we have been working with ESD competences over the past four years, we have seen many positive outcomes for our undergraduates. These include increased knowledge, growing self-confidence and a range of positive actions that they have taken. Given that we intend ESD competences to help bring about

P. Vare et al. (eds.), *Competences in Education for Sustainable Development*, Sustainable Development Goals Series, https://doi.org/10.1007/978-3-030-91055-6_21

individual—and ultimately societal—change, we feel this is a good start.

One aspect that we have wrestled with, however, is how to assess these outcomes effectively. As Kerry Shephard illustrates in Chap. 6, the assessment of ESD competences is not as straightforward as might first appear. The whole point of ESD is to stimulate some form of intrinsic motivation to make more sustainable life choices but as soon as we confer credit for demonstrating such outcomes, we immediately introduce a strong extrinsic motivation for students to claim to have made sustainable life choices in a performative manner.

We are aware that we have been tackling this in different ways so we are interested to learn about the approaches you have taken. We, Paul and Rick, know that you worry about the idea that assessment conveys the notion of measurement which you feel is inappropriate. Can you explain why?

Michela/Francesca One thing we should be clear about from the outset is that we need to look for an assessment process which takes account of the *complexity* of ESD and explores the *quality* of the processes and transformations implemented.

The fact that we may define criteria and assign numbers to the observed outcomes and by consequence can order them according to a numerical scale, does not make this a measurement. For instance, Intelligence Quotient (IQ) is not a 'measurement', even if it is still in use despite much criticism (e.g. Gould 1981), just as the Michelin stars of restaurants or the ratings expressed on websites are not measurements.

The paradigm that should inspire educational research and practice cannot be the 'Galilean' or 'Newtonian' one which seeks to simplify and quantify the complexity of the real world in order to establish objective relationships, but rather one like that proposed by Ginzburg (1989) for human sciences: an 'evidential' or 'circumstantial' paradigm ('*paradigma indiziario*'). In this paradigm, small differences and small signs enable the historian, the psychologist, the investigator and the educator, to rationally reconstruct and understand what has probably happened. The aim is not to find simple general rules or to collect defined outcomes but to reconstruct transformation histories that are intrinsically unique.

The consideration of educational evaluation as a measurement has been criticised several times in the past. For example, within environmental education Flogaitis and Liriakou (2000), following a proposal of Robottom and Hart (1993), denounced the predominance in educational evaluation of a positivistic paradigm and proposed a socio-critical paradigm where reality is conceived as a complex matter, knowledge is socially constructed and evaluation is one of the instruments of change. Here, the evaluator is viewed as a social agent of change and uses their judgement, based on stated and shared criteria, to support the transformation process.

Aaron Exactly! There are serious pitfalls when tackling assessment but, in my view, this argues for investing in it rather than ignoring it and the time is ripe for making a serious effort to develop legitimate assessment processes for ESD. Not only is there agreement about the goal of ESD being to support sustainability transformations (Franco et al. 2019), but there is an increasing convergence around specifying what the learning objectives should be for the students who are to be educated in facilitating these transformations (Brundiers et al. 2021; see also Chaps. 3 and 4 of this book). This is important progress and if a university wished to start a degree programme in ESD, they would have a solidly informed base from which to articulate its intended outcomes. However, if they were to ask HOW they should structure their curriculum or their teaching in order to achieve these outcomes, the field of ESD research would have little empirical evidence to offer.

There are certainly plenty of case studies about exciting and interesting programmes, courses and university efforts (Weiss and Barth 2019), but these remain largely descriptive. The intention of the Educating Future Change Agents (EFCA) project (Redman 2020) was to move beyond this and utilise cases to build empirical evidence about how we can achieve competencies in sustainability. It was immediately apparent that a critical component of that effort was going to be properly and rigorously assessing competencies. Such assessments would indicate whether certain teaching approaches, or curriculum structures, were more or less effective and make empirical comparisons between cases possible. Yet when we looked into the existing research, it offered little guidance and relied heavily on students' self-assessment of their own competence, an approach whose weaknesses are well catalogued (Redman et al. 2021).

As Michela and Francesca point out though, the last thing that we want to fall into is simplifying ESD down to something that can be assessed with a standardised multiple-choice test. Fortunately, one of the advantages of using a competency-based approach for learning objectives is that it preserves that real-world complexity in a way that foils the traditional modes of assessment (Frey and Hartig 2009). Yet, neither can we throw our hands up and say they are impossible to assess. The field of ESD has long argued that novel teaching approaches are vital (Frisk and Larson 2011), yet if we cannot provide evidence (via assessment) that these methods are achieving their stated goals of developing sustainability change agents who can facilitate transformations, then we should (rightfully) expect our calls for these innovative pedagogies to be increasingly ignored.

Paul/Rick It seems that we all feel similarly that assessment of competence in ESD is important so that we can provide evidence of the effectiveness of our teaching. However, the challenge is how to do that in a way that is constructively aligned with our defined outcomes and pedagogical approach.

This notion of transformation complicates matters. If the ultimate aim of ESD is the transformation of society, then presumably assessment should be assessing the non-linear interactions that would need to take place over an extended period of time to see if a given programme of educator preparation had led to corresponding classroom actions, which in turn led to their students adopting positive behaviours and attitudes commensurate with a more sustainable society.

Clearly it is not feasible to assess this whole process, in which case we have to determine what we *can* look for and decide whether that can provide sufficient evidence to suggest that our teacher education has made this transformation more likely to occur. In other words, we are seeking specific ingredients that, if evidence can be found for them, would convey the *likelihood* of transformation emerging, possibly over time, at the level of future learners and eventually at system level.

Given that our work is focused on training the educators, then presumably we need to look for this evidence in, or from, them. Demonstration of ESD competences however, is unlikely to be sufficient because they may have *competency* (the ability), but not the *competence* (to put it into practice).

A broad concept of competence (see the discussion in the Introduction of this book) suggests more than just ability, it also encompasses the values that would lead to the motivation to apply their learning and the *agency* to be able to action them. On this last point, Campbell (2009) identifies two types of agency:

Type 1: the ability to operate freely at the individual level, albeit within existing structures. Campbell terms this the *power of agency;* we might also call this competency.

Type 2: the ability and confidence to make changes in the face of structures thereby contributing to societal change—what Campbell terms *agentic power*; we would see this as fully operational competence.

Given the constraints that bind many educators, e.g. operating within prescribed curricula and tightly controlled standardised assessments, the options for Type 2 agency seem limited, yet this is exactly what is needed, a willingness to find—or create—the 'wiggle room' required to open up possibilities for implementing their own ESD competences.

To add further complication, if we truly wish to encourage critical thought, autonomy and agency, then we cannot control how this agency will be used and where the critical thinking might lead; indeed, critical thinking has its own dangers (see Chap. 7). As a consequence, it feels like the best we can hope for is to take the 'evidential' approach, as suggested by Michela and Francesca, and look for evidence or indicators of (a) competency, that is the *ability* to do things as an individual and (b) competence, which for us includes the *willingness* of student teachers and educators to find opportunities to demonstrate this ability despite the constraints of their professional context.

If we can find evidence of these elements, then perhaps we can infer from this that the ingredients necessary to achieve societal change are present *insofar* as we are able to instil them. The extent to which these changes actually occur *and* contribute towards a more sustainable future will always be determined by the agency of individuals who are subject to a variety of often unforeseen constraints and influences at personal, professional and societal levels.

Trusting this clarifies what we are looking for, we are left with the question of how we find that evidence.

Aaron, you have researched various ways of assessing competences, what did you discover?

Aaron Our systematic review of the literature revealed that the body of research on assessing competences in sustainability has grown rapidly in recent years (Redman et al. 2021). Yet despite this growth, the field is clearly still in its infancy and offers little empirical guidance for either practitioners or researchers interested in effective assessment. There are several ways in which current practice (at least as evidenced in the literature) is hobbled. The first is an underinvestment in development of tools for assessment. This manifests in individual studies where assessment serves merely to produce data about some kind of pedagogical or curricular innovation, as well as in the fact that there are few instances where research groups are building on each other's (or even their own) work. Perhaps driven by this underinvestment, the most widely used tool is the weakest, scaled self-assessment, used in well over half of the studies.

But secondly, more fundamentally what we saw hampering effective assessment was touched on by Paul and Rick, which is properly aligning assessment with the desired outcomes of ESD. As they pointed out, the outcome of leading transformations is too ambitious to possibly capture in one (or many) assessments (if possible at all). Currently, this challenge is hand-waved away with limited assessments being used to make broad statements about competence development. However, what is needed is to be explicit about what specific pieces of the overall outcome you are intending to assess and then utilise the approaches which give you the best evidence about those specific pieces. While dispersed, the initial indications of what tools might be best in what circumstances and to measure what components of competence, do exist in the current research.

The typology of tools, which we distilled from the literature, brings together the findings of the field to enable one to evaluate options when selecting assessment tools. We identified eight distinct types of tools: scaled self-assessment, reflective writing, scenario/case test, focus group/interview, performance observation, concept mapping, conventional test and regular course work. These can be clustered into three meta-types, namely self-perceiving-based, observation-based and test-based assessment procedures. This typology provides a framework on which we can layer more findings, explore additional tools and identify the best assessment approach for our specific purposes.

Michela/Francesca This range of assessment approaches is interesting for us and we have actually used many of them for assessing the learning process in our context. However, the identification of appropriate assessment tools, as Aaron states, strongly depends on the main purpose(s) we aim to achieve with our educational project and on what type of change we expect to stimulate through it.

Therefore, we probably need to start again from the initial question posed by Paul and Rick: WHAT are we aiming to assess? In our context the aim of the assessment was to evaluate the training of educators as change agents by looking for the demonstration of competences in their professional contexts, as Paul and Rick have underlined and as we have explored in detail in Chap. 11.

An important goal in our training programmes was that our learners became more attentive to the complexity of the world and of the educational processes, became competent (in the sense of competency), but were also willing to use their competences to address future uncertainties.

Now, returning to the question posed by Paul and Rick about HOW to assess competences and how to find evidence, the key, for us, is to look at the range of tools and approaches introduced by Aaron not as separate, but interwoven: if self-assessment is fundamental, for example, to look at the increase in consciousness, it is the interweaving with other tools, i.e. focus groups and peer evaluation of reflective writing of experiences, which can provide a more complete and 'three-dimensional picture' of what we are trying to assess.

Another important aspect to consider in relation to how to find evidence is the necessity to focus attention not only on *expected* results, but also on *emerging, unforeseen* outcomes in order to detect the changes instilled by the educational process which is by nature complex and dynamic.

Consequently, in our training programme in Italy (see Chap. 11) we tried to 'follow the transformation' and understand: (1) if we were successful in promoting in our learners a change in their vision and beliefs about ESD and the educators' role, and if so (2) in which direction and (3) with what level of consciousness this was occurring. This is because we think that the willingness to put competences into practice cannot be generated without an increase in the awareness of being change agents.

To answer these three questions, we had to follow the transformation process from its inception while remaining open to the detection of unforeseen elements and signals. All this had to align with the specific competence framework's underlying values and concepts (Farioli and Mayer 2020).

In order to gather the required evidence, we used an 'environmental autobiography' tool at the start of each training programme. This told us how learners felt about their role as sustainability educators as well as about their emotions, willingness and potentialities to engage as change agents.

It was only by knowing learners' starting points that we could understand, by the end of the course, the change that had actually occurred and the extent to which this could be attributed to the course itself, rather than being an outcome of the competences and knowledge that the learners already possessed.

Thereafter, the use of interwoven assessment tools, i.e. observation of assigned tasks, analysis and peer assessment of reflective writing and construction of individual portfolios that mapped evidence of experiences and competences achieved at different stages in the process, has been crucial for us in order to 'follow the transformation'.

Each of the tools that were used captured only some aspects of the learning, but it was the interlacing of the results that provided us with a more accurate and complete picture of the changes that were taking place. For example, a storytelling analysis exercise, carried out in groups, was fruitfully piloted in connection with peer assessment and focus groups and allowed each of our learners to 'look at themselves from the point of view of others' and to 'reflect into others'.

The iterative process of practice (in the sense of carrying out assigned tasks during the course), reflection in action and challenge by others has—and this is for us the most important result we have achieved—favoured a path of consciousness in our learners of the competences that they have developed and acted on and of those that they have yet to develop, improve and put into practice.

The challenge however is in how to tie together the clues and evidence which emerge from the different assessments in place and how to interweave them in order to build a consistent framework for an overall assessment with a rounder sense of purpose. Such a framework should not aim to be an 'objective' assessment, since it is never 'neutral'; even in test-based procedures such as the Programme for International Student Assessment (PISA), subjectivity is always present, for example in the selection, however negotiated, of the questions asked. However, as our experience has demonstrated, it should aim to be useful for learners, allowing them to look at themselves and their professional path with new eyes and to feel more confident in their acquired competences, all of which will probably render them more capable of instigating change.

Quite which clues and indicators are to be used for evidence, and HOW to best combine them for a meaningful and appropriate assessment approach, is the key to a 'quality assessment', as well as for assessing the 'quality of change'. In our opinion, this remains one of the issues in ESD research that requires further investigation and empirical evidence.

Paul/Rick What has become apparent from these contributions is that no single assessment tool can capture the complexity of what we are aiming to assess and that there is no perfect solution. In our own work, we have also relied on a mixed methods approach.

In our context, student teachers need to demonstrate the achievement of specific academic standards in order to be awarded credit *as well as* develop competences. However, we have been fortunate to run a non-credit bearing, competence-based programme for four years and this provided the opportunity to develop our assessment approach before extending the programme to accredited courses.

We asked participants to keep a reflective journal throughout the programme outlining how they had applied the competences in their professional, social and/or private life and, where applicable, how they had helped develop the competences in others. A thematic analysis of these journals provided evidence of three key outcomes:

- Understanding of the competences and the issues they raise
- Action taken on the basis of the competences
- Reflection on the competences themselves and on their own engagement with them

Each of these outcomes were broken down further into three sub-categories or 'learning aspects', which we have listed elsewhere (Vare and Millican 2020). In our case, we were working with the twelve Rounder Sense of Purpose (RSP) competences and recognised that seeking evidence of nine learning aspects for each competence would be unrealistic and would sacrifice depth of engagement for breadth of coverage. We decided that a meaningful indicator of the *extent* of a student's learning across all twelve competences would be if they provided evidence of at least four of the nine learning aspects under each competence, with at least one in each category (Understanding, Action & Reflection). We also sought evidence of each of the nine aspects in at least four competences.

Unsurprisingly, analysis of students' reflective journals revealed qualitative differences in the *depth* of engagement or levels of ability. Using exemplar statements from the journals we drafted descriptors for different levels of achievement in relation to the nine aspects of learning. This enabled us to create a marking grid similar to those used by colleagues on our other accredited courses. By shading the 'best fit' descriptors for each aspect of learning, an assessor builds an impression of a learner's competence; this allows for a composite grade to be reached thus fulfilling university requirements.

Used together, these two tools can be used to assess a range of evidence including reflective journals, videos and formal essays. While any form of assessment will give an incomplete picture and be based on the professional judgement of the assessor, we hope that this balance between *extent* (quantity) and *depth* (quality) can go some way towards assessing competence and competency as well as an indication of transformation with its promise of sustained change.

Aaron As already mentioned, for the EFCA project we found a need to develop our own assessment approach and ultimately deployed ten different tools which spanned the whole range of types described earlier. Similar to Paul/Rick and Francesca/Michela, we were able to make the most robust assessments of students' competencies by triangulating results of different assessments in cluster 1 (student self-perception). One particularly strong approach was to ask students to rate their level of competency and write a short justification (Birdman et al. 2021) which was then used as a starting point for interviews. This process was repeated four times throughout a two-year degree programme and gave a robust overview of the students' individual development, but did not enable a comparative 'measurement' or empirical comparison between students.

Two studies also attempted to construct domain-specific, yet holistic, test-based assessments and significant time was invested in developing and piloting these tools. One of the tests used real-world curriculum and expert judgement to assess the students' Pedagogical Content Knowledge (Brandt et al. 2019), while for the other a full in vivo simulation was run inspired by advanced approaches taken in medicine (Howley 2004). In this live simulation, students were confronted with a mock city council to which they had to offer their advice on the sustainability of an economic development plan (Foucrier 2020). These assessments gave insight into the competency development of the group, but little in terms of tracking individuals. They also suffered from the fact that they were not 'graded', which certainly influenced student effort. As other studies have found, variance in scores may be largely driven by effort invested (Zamarro et al. 2019).

In conclusion, our three cases give an example of the rich variation of assessment approaches being taken with ESD, but highlight the critical need for a more comprehensive and coordinated approach to be taken. The experience at EFCA was that even with significant resources and explicit support from both institution and instructors, it was not possible to administer a consistent and robust set of assessments across its studies. This chapter therefore serves as a starting point for the necessary conversation between academics and practitioners to both learn from, and build upon, each other's work in order to develop holistic and effective approaches to assessing students' development of competency, so that they can be effectively supported to become the change agents that the world needs.

References

Birdman, J., Redman, A., & Lang, D. J. (2021). Pushing the boundaries: experience-based learning in early phases of graduate sustainability curricula. *International Journal of Sustainability in Higher Education*, 22(1), 237–253. https://doi.org/10.1108/IJSHE-08-2019-0242

Brandt, J. O., Bürgener, L., Barth, M., & Redman, A. (2019). Becoming a competent teacher in education for sustainable development: Learning outcomes and processes in teacher education. *International Journal of Sustainability in Higher Education*, 20(4), 630–653. https://doi.org/10.1108/IJSHE-10-2018-0183

Brundiers, K., Barth, M., Cebrián, G., Cohen, M., Diaz, L., Doucette-Remington, S., Dripps, W., Habron, G., Harré, N., Jarchow, M., Losch, K., Michel, J., Mochizuki, Y., Rieckmann, M., Parnell, R., Walker, P., & Zint, M. (2021). Key competencies in sustainability in higher education—toward an agreed-upon reference framework. *Sustainability Science*, 16(1), 13–29. https://doi.org/10.1007/s11625-020-00838-2

Campbell, C. (2009). Distinguishing the Power of Agency from Agentic Power: A Note on Weber and the "Black Box" of Personal Agency. *Sociological Theory*. 2009;27(4):407-418. https://doi.org/10.1111/j.1467-9558.2009.01355.x

Farioli, F. & Mayer, M. (2020), *Transformative Evaluation for Sustainability Educators: the experience of a Competence-based Educational Research project*, 2nd GUNI International Conference on SDGs: Higher Education & Science Take Action, Summary Report, p.101 - 104

Flogaitis, E. & Liriakou, G. (2000). Quelle évaluation pour quelle Education relative à l'Environnement? [What Evaluation for What Environmental

Education?] Education relative à l'Environnement, *RegardRechercheRéflexions*, 2, pp. 13–30

Foucrier, T. (2020). *Training future entrepreneurs – developing and assessing sustainability competencies in entrepreneurship education* (Order No. 27955218). Available from Dissertations & Theses @ Arizona State University; ProQuest Dissertations & Theses Global. (2406963829)

Franco, I., Saito, O., Vaughter, P., Whereat, J., Kanie, N., & Takemoto, K. (2019). Higher education for sustainable development: actioning the global goals in policy, curriculum and practice. *Sustainability Science*, 14(6), 1621–1642. https://doi.org/10.1007/s11625-018-0628-4

Frey, A., & Hartig, J. (2009). Assessment of competencies. *Studies in Educational Evaluation*, 35(2–3), 55–56. https://doi.org/10.1016/j.stueduc.2009.10.001

Frisk, E., & Larson, K. L. (2011). Educating for Sustainability: Competencies & Practices for Transformative Action. *Journal of Sustainability Education*, 2 (March).

Ginzburg C. (1989). Clues: Roots of an Evidential paradigm in *Clues, Myths, and the Historical Method*, The John Hopkins University Press

Gould, S. J., (1981) *The Mismeasure of Man*. New York: W. W. Norton & Company.

Howley, L. D. (2004). Performance assessment in medical education: Where we've been and where we're going. In *Evaluation and the Health Professions*. https://doi.org/10.1177/0163278704267044

Redman, A. (2020). *Assessing the Development of Key Competencies in Sustainability* [Arizona State University]. In ProQuest Dissertations and Theses. http://login.ezproxy1.lib.asu.edu/login?url=https://www.proquest.com/docview/2445352259?accountid=4485

Redman, A., Wiek, A., & Barth, M. (2021). Current practice of assessing students' sustainability competencies: a review of tools. *Sustainability Science*, 16(1), 117–135. https://doi.org/10.1007/s11625-020-00855-1

Robottom, J. & Hart, P. (1993). *Research in environmental education. Engaging the debate*. Victoria, Deakin University

Vare, P and Millican, R (2020) Educating for a 'Rounder Sense of Purpose' - Lessons learned from practice: assessing ESD competences. Presentation to TEESNet Conference, 2020. Accessed January 2021 at: https://eprints.glos.ac.uk/id/eprint/10434

Weiss, M., & Barth, M. (2019). Global research landscape of sustainability curricula implementation in higher education. *International Journal of Sustainability in Higher Education*, 20(4), 570–589. https://doi.org/10.1108/IJSHE-10-2018-0190

Zamarro, G., Hitt, C., & Mendez, I. (2019). When students don't care: Reexamining international differences in achievement and student effort. *Journal of Human Capital*. https://doi.org/10.1086/705799

Part IV

Conclusions

22 Lessons Learned and Future Research Directions in Educating for Sustainability Competencies

Ingrid Mulà, Gisela Cebrián, and Mercè Junyent

Abstract

Transformation towards sustainable development requires people who are motivated and capable of challenging current systems. But what competencies are needed to facilitate and implement effective change for sustainability? This question has been central for the past decade to scholars interested in sustainability and education for sustainable development leading to the development of generic sustainability competence frameworks. As the field of sustainability competencies is consolidating, important criticisms are raised about the lack of conceptual clarity of the competencies proposed and how these can be developed, supported, and assessed. In addition, the discourse has been dominated by North American and European perspectives leading to a cultural bias in the definition and interpretation of these competencies. Also, considering how social and institutional structures can hinder or facilitate the development of people's capacities in relation to sustainability, little attention has been paid to the need to contextualize competencies within sociocultural and institutional settings. Furthermore, with COVID-19 accelerating and mainstreaming e-learning, challenges are presented in terms of what online pedagogies can be used to support the acquisition of these competencies. In this chapter, we capture some lessons learned from recent work and suggest some future directions in order to instigate new developments in this area.

Keywords

Competencies · Capabilities · Skills · Education for sustainable development · Sustainability

Introduction

Transformation towards sustainable development requires capable people 'willing to' challenge the *status quo* (Shephard et al. 2019), as well as to engage in different ways of thinking, acting, and living (Sterling 2001). A broad agreement exists that education and learning can help us to explore alternative lifestyles (UNESCO 2004; Tilbury 2011), but there is also an acknowledgement that

I. Mulà (✉)
Institute of Educational Research, Universitat de Girona, Girona, Spain
e-mail: ingrid.mula@udg.edu

G. Cebrián
Department of Pedagogy, Universitat Rovira i Virgili, Tarragona, Spain

M. Junyent
Research Group COMPLEX, Universitat Autònoma de Barcelona, Barcelona, Spain

P. Vare et al. (eds.), *Competences in Education for Sustainable Development*, Sustainable Development Goals Series, https://doi.org/10.1007/978-3-030-91055-6_22

what we have learned to date does not prepare us for the challenge of sustainable development (UNESCO 2020). While most of the official formal education curricula aim at empowering citizens to become critical, empathic and active change agents towards more sustainable, socially just and equitable societies, implementation remains a challenge (Cebrián et al. 2020).

Signs of progress have emerged during the past years, demonstrating that a change in education is possible. From early childhood to higher education, theoretical models and practical efforts to embed Education for Sustainable Development (ESD) through whole-school approaches have been documented (UNESCO 2014). Evidence also exists showing that teaching and learning methodologies are changing. Knowledge transmission pedagogies are being challenged across the different formal education levels and are slowly being replaced by active methodologies supporting more student-centered, participatory and democratic learning processes (Sterling 2001; Tilbury 2011; Lozano et al. 2017). However, these practices remain niches and far from mainstream efforts (Ryan and Tilbury 2013) and, often, although innovations are introduced, the values underlying the education process are untouched. Education reforms are no longer useful if we want learners to engage in transformative processes that fundamentally challenge their thoughts, feelings, and actions (O'Sullivan et al. 2002; O'Sullivan 2003). These transformative experiences require a fundamental questioning of the purpose of education and the role of educational institutions (Sterling 2001; UNESCO 2015), as well as disrupting learning approaches by using more critical, emancipatory and relational pedagogies (Wals 2020).

We agree with Sterling et al. (2017) that sustainability competencies can be a starting point to leverage pedagogical transformation and stimulate fundamental systemic changes in educational organizations. Competence-based approaches are also aligned with the ambitions of Agenda 2030 (target 4.7 calls for supporting learners "to acquire knowledge and skills needed to promote sustainable development") (UN 2015) and UNESCO's vision spelled out in its "ESD for 2030" Framework and Roadmap (UN 2019; UNESCO 2020). Planning with sustainability competencies in mind leads us to ask critical questions about what, where and how we learn, as well as to assess whether learners can fully develop as human beings and contribute to creating more attractive sustainable futures for all.

The last decade has seen increasing research interest in defining what 'knowledge, capacities and skills, motives and affective dispositions' are needed to facilitate societal transformation (Rieckmann 2012, p. 129). Generic sustainability competence frameworks have been proposed (Rieckmann 2012; UNESCO 2017), together with more specific frameworks for schoolchildren (de Haan 2006), higher education students (Brundiers et al. 2021; Lozano et al. 2017; Wiek et al. 2016; Wiek et al. 2011), sustainability entrepreneurs (Mindt and Rieckmann 2017), or educators (Sleurs 2008; Strachan 2012; UNECE 2012; Bertschy et al. 2013; Rauch and Steiner 2013; Cebrián and Junyent 2015; Vare et al. 2019).

In relation to educators, the existing frameworks have focused on defining the ESD competencies that should be developed through teacher education (Bertschy et al. 2013; Cebrián and Junyent 2015; Rauch and Steiner 2013; Sleurs 2008), and on the ESD competencies that educators from any field and education level should possess (UNECE 2012; Vare et al. 2019). All these frameworks are concerned with educators' abilities and behaviors while the ones focused on teacher education tend to include differentiations between teachers' individual ESD competencies and the type of learning and competencies that teachers should promote within the school context as members of the school.

One criticism of the recent literature is the lack of conceptual clarity and rich description of the competencies that have been proposed (Glasser and Hirsh 2016; Sterling et al. 2017). Designing competence-based education requires clear pedagogical and assessment strategies on how learners develop and are willing to use these competencies to contribute to a collective social transformation of our societies towards sustainability. In addition, competencies must be appropriately contextualized within a sociocultural and institutional setting, considering how the social and institutional structures can hinder or facilitate the development of people's capacities in the area of sustainability.

In this chapter, we capture some lessons learned from the work carried out in the past years and suggest some future directions in order to instigate new developments in this area.

Conceptual Terminology and Rich Descriptions of Competencies Are Needed

The focus on competencies in the field of sustainability education has attracted attention from academics, policy-makers, and practitioners across the world from a diversity of backgrounds and cultural contexts, leading to several instances of terminological and conceptual confusion. Sterling et al. (2017) argue that to move forward further clarification and appropriate use of terms are needed. Words with subtle nuances like 'competences', 'competencies', 'capabilities', 'attributes', or 'generic skills' are often used in the literature equivalently. Around the world, some countries prefer to use specific terms over others and, in other countries, some of the subtleties in meaning among terms do not exist. For example, in Australia, 'capability' is more frequently used than competence or competency, and in the UK there is a preference towards the use of 'skills'; in Spain, there are no differences between 'competence' and 'competency' which are both translated as *competencia*. In this chapter, we understand sustainability 'competence' as the overarching term that refers to the knowledge, skills, attitudes, and values necessary to effectively perform tasks, solve real-world sustainability challenges, and support the transformation of processes and systems. We also understand that this competence can be broken down into a set of different "competencies" (in singular, "competency").

In addition, in some of the work published there is no specific distinction between 'sustainability' and 'ESD' competencies, leading to another level of academic misunderstanding. In our view, sustainability competencies refer to the knowledge, skills, attitudes, and values that enable one to act in a sustainable manner in real-world situations (Wiek et al. 2011), while ESD competencies refer to educators' competencies to foster sustainability competencies through ESD processes and practices (Vare et al. 2019).

Although the competencies' approach is popular across the globe, there are many scholars that have criticized its discourse as being instrumental, utilitarian, and market-oriented (Edwards 2016). Some authors, like Lotz-Sisitka et al. (2015), suggest that adopting the capabilities approach (Sen 1993) might be more useful in order to re-think learning and pedagogical development. Lozano et al. (2012) stress that the capability approach is more interested in people's values, freedom, and agency, as opposed to the competencies approach that has a stronger focus on solving concrete problems of specific demands. This resonates with the idea of O'Donoghue et al. (2007) that sustainability should be considered a challenge to be fully taken on, rather than a problem that needs to be solved.

We acknowledge that the capability approach is powerful due to its value-based orientation, but also endorse the decision of the experts participating in the study conducted by Brundiers et al. (2021) to retain the word 'competency' in view of its broad use in the context of education and sustainable development. In addition, in their philosophical hermeneutic analysis of ESD papers, Shephard et al. (2019) remind us that the concept of competence, as defined by the ESD community, has always been centered on values and been underpinned by ideas related to freedom of choice and learners' self-determination. An important conclusion of their work is that whether using one or another approach, we cannot assume that those who have developed sustainability competencies decide to use them in every context. Therefore, it becomes fundamental to engage children from early ages to understanding the need for change and support them throughout life, in different ways and through different strategies, to develop and use their sustainability knowledge and skills in all possible situations.

Curriculum Design and Development Must Support Education Around Sustainability Competencies

Much of the current debate on sustainability competencies revolves around the design of curricula and educational programs. Glasser and Hirsh (2016) point out that consensus is needed on, first, what competencies must be fostered and, second, how these competencies can best be supported and assessed. Most of the frameworks defined so far have been developed based on literature reviews and expert opinions and, with some exceptions, presented as lists of complex ideas (Wilhelm et al. 2019) difficult to achieve and assess (Brundiers et al. 2021). Timm and Barth (2021) stress that only recently have the first research projects examined empirical evidence regarding how sustainability competencies contribute to both successful teaching and implementation of ESD, thus illustrating the need to accelerate research in this area.

With the aim of connecting competencies with pedagogical approaches, Lozano et al. (2019) conducted a research survey with European Higher Education lecturers (see also Chap. 17). The authors concluded that competencies must be supported by a combination of methods and, when using more traditional approaches (such as lectures), educators should reflect on how best they can support the development of the competencies. As a contribution, this paper presents a matrix that connects what methods can be more effective to teach the different competencies identified. Although a valuable study, the findings must be interpreted openly so that the critical creativity of educators is not hindered through the use of 'another list', this time of teaching and learning methods.

The existing literature is full of studies that have documented how methods like problem- or inquiry-based learning have been used to foster sustainability competencies (Thomas 2009; Wiek et al. 2014). Although the results of their implementation are positive in terms of competence development, it seems that these efforts are not really having the transformative effect that we seek in our societies. Recent research into transformative learning for sustainability calls for the development of more hybrid and engaged pedagogies involving multiple actors and voices (Lotz-Sisitka et al. 2015; Wals 2007), the creation of open and transdisciplinary learning spaces (Bürgener and Barth 2018), and the exploration of the dialectics between cultural tradition and innovation (Tilbury and Mulà 2009; Tilbury 2011). This requires building bridges between formal, non-formal and informal learning, creating social learning spaces to confront diversity of values and ideas, challenging the role of and relationships between educators, learners and stakeholders and rethinking assessment of learning. In other words, it implies challenging how the learning process and the curriculum are traditionally designed (especially challenging Eurocentric perspectives) and reconsidering how sustainability competencies can best be fostered.

A significant challenge described in the literature is finding appropriate paths to assess student sustainability competencies (Cebrián et al. 2020). To start with, we believe it is important to consider Sadler's (2013) point about the risks of decomposing competence into a set of competencies—a common practice in the sustainability field. While defining a set of manageable competencies simplifies the assessment exercise and facilitates judgements of whether or not each competency has been achieved, it obscures how an individual performs the different competencies together as a whole (see also discussion on "the RSP palette," Chaps. 2 and 4). For Sadler, essential to the assessment of competencies is how students are able to 'orchestrate' them independently and proficiently within different contexts (p. 11). Analyzing the whole (*is the student competent in relation to sustainability?*), and not only the parts (*has the student fostered each of the different sustainability competencies defined?*), has certain benefits. Firstly, it solves the problem related to defining the boundaries between competencies. There is an inevitable overlap of ideas and principles among the different sustainability competencies proposed, which makes it difficult to assess competencies as separate blocks (for example, it is difficult to perform a strategic competency without performing a future thinking competency). Second, seeing the whole and not the parts in isolation opens a window for greater criticality and creativity in the assessment process. It provides the opportunity for learners to perform other competencies (not pre-defined in the assessment exercise) which can be instrumental in a particular context.

The field of assessment of sustainability competencies is developing slowly and most experiences documented so far tend to focus on assessing individual competencies, failing to demonstrate, with rich descriptions, how learners perform them together as a whole. A study recently published by Redman et al. (2021) provides a picture of the different tools utilized to assess learners' sustainability competencies (see Chap. 21). The authors propose a typology of eight assessment tools divided into three groups: (1) self-perceiving (scaled self-assessment, reflective writing, and focus group/interview); (2) observation (performance observation, regular course work, and conceptual mapping); and (3) test-based approaches (scenario/case test and conventional test). Surprisingly, and responding to Mogensen and Schack's (2010) calls for particular attention to self-evaluation, self-assessment methods were disproportionally represented among the articles reviewed. This could be explained as being in the context of summative rather than (trans)formative assessment. Cebrián et al. (2020) argue that more research is needed on the latter to support student learning in more meaningful and effective ways (Black and William 1998) and to guide educators to do a better job (Popham 2008).

Finally, it is worth mentioning the work of Holdsworth et al. (2020) as it offers another approach to assess sustainability competencies in higher education. The authors present a framework to explore how higher education graduate sustainability capabilities are applied in professional settings in the workplace after graduation. Although a complex and tedious approach to implement, it provides educators with useful data that can show whether graduates are applying the competencies in real contexts. Further research in this direction, also exploring how competencies are used in personal and community contexts (and not only professional settings), can further support the process of designing and facilitating learning for sustainability processes.

The Context in Which Sustainability Competencies are Defined and Developed Is Important

One does not need to undertake a rigorous systematic review to realize that the sustainability competence discourse is dominated by North American and European perspectives. In addition to the terminological confusion and the lack of consensus on what competencies are needed, scholars have recognized that there is an obvious cultural bias in the sustainability competence frameworks available in the literature (Rieckmann 2013; Bürgener and Barth 2018; Brundiers et al. 2021). The review carried out by Sterling et al. (2017) also shows that the majority of articles published in this area refer to higher education contexts, with only a few publications addressing early childhood, primary, secondary, or adult education. There are practically no experiences that refer to informal and community-based forms of learning.

Some examples of work have contributed to include different cultural voices in the definition of sustainability competences. An example is Rieckmann (2012) PhD thesis that presents a joint discourse of European and Latin American experts. In his study, minor differences are revealed between participants from both parts of the world. Europeans put more emphasis on the need of competencies related to 'empathy' and 'change of perspective' and Latin Americans on 'cooperation' and 'participation'. Another example is the study developed by Demssie et al. (2019) who question whether sustainability competencies proposed to date are universally relevant; they offer an Ethiopian and "base of the pyramid" perspective. Involving 33 experts from academia and industry, the authors conclude that several competencies such as 'systems thinking' may be considered universal, whereas others such as 'competence to utilize indigenous resources for sustainability' could be considered context specific. In another study, the same authors (Demssie et al. 2020) explore opportunities to embed indigenous knowledge systems in mainstream modern (westernized) education in Ethiopia with a view to developing sustainability competencies. The results highlight that using indigenous learning approaches requires more open, collaborative and community-based pedagogies, aligned with transformative learning approaches. Along the same lines, Dai and Hwang (2019) carried out empirical research on bamboo crafting courses in universities and determined that knowledge and skills learned are better brought into play when contextualized in social practice which, in turn, helps students to develop cultural self-confidence.

It is also important to note that certain competencies that are likely to be considered universally relevant (e.g., critical thinking) might be interpreted differently in different sociocultural contexts. Rather than seeing this as a problem, we should use it to enhance intercultural dialogue on sustainable development and enrich our own cultural understandings. As pointed out by Yoneyama (2012) and other authors of post-colonial literature, the often-claimed perceived weaknesses of certain competencies among people from a particular culture create a divide between regions and cultures. This leads to seeing people from other cultures as "the other," rather than embracing the richness that every individual can bring to the transition towards sustainability. Tilbury and Mulà (2009) state that intercultural dialogue is central to sustainable development as it implies understanding, respecting and forging links among cultures, as well as exchanging and co-creating knowledge to seek and re-invent more creative ways to live together.

In order to cultivate sustainability competencies, there must also be a social and institutional environment that allows them to occur. Following up with the example of critical thinking, critical pedagogies are commonly used to engage students in debating provocative and sometimes uncomfortable issues, as well as to empower and support them in order to bring about social justice and transformation (McLaren 1994). This requires a social context that helps people to engage in these types of debates freely, but also institutional structures and educational systems that encourage and reward educators who use this critical approach. The positive side is that a focus on sustainability competencies can, at the same time, influence the social context in which they are operationalized. As Sterling et al. (2017) argue, competencies are a vehicle to catalyze pedagogic transformation, institutional learning, and structural change, speeding up the process of embedding sustainability institutionally. We must, thus, interpret the competence approach as a more complex endeavor than just supporting individuals' knowledge, skills, and attitudes, since it also represents an effort to transform broader systems towards sustainability.

Finally, competencies might be expressed differently depending on the discipline and knowledge areas, as well as on the different educational forms and levels (Sadler 2013). Therefore, even if we managed to reach a consensus on what sustainability competencies are needed, rich descriptions that contextualize them in the particular cultural and educational setting where they are operationalized are essential.

Further Research and Experimentation Are Needed to Explore How Sustainability Competencies Can Best Be Developed Through Online Learning

Before COVID-19, there was already high adoption of e-learning and educational technologies across the different education areas and levels (Lim et al. 2013; Panigrahi et al. 2018), pointing to the urgent need to explore the implications in relation to learning for sustainability and the development of sustainability competencies. The pandemic has forced everyone to move to online learning. While some think that we should move away from this due to its negative impacts on equity and quality as well as due to the ecological impact of digitalization, others believe that a new hybrid model of education will materialize, bringing more benefits in comparison to the old one (Hohlfeld et al. 2017; Zhang et al. 2020).

The value of online learning for sustainable development has been recognized internationally (Leicht et al. 2018; UNESCO 2014). Among others, e-learning shows the potential to reach students who otherwise would not be able to participate in person and in real-time education, supports international, intercultural, and intergenerational dialogue on sustainability topics and provides access to a wide range of resources (Ally 2008). However, a review on Massive Open Online Courses (MOOCs) on climate change showed that there is an emphasis on knowledge transfer approaches and a lack of adequate pedagogical mechanisms to contextualize transformative action learning and assess sustainability learning outcomes (Lotz-Sisitka 2014). Thus, after many years and efforts to support active, participatory and social learning approaches, there is a risk of going back to where we were before.

There are several challenges that the sustainability educator faces when designing online courses from scratch or adapting programs that used to be facilitated face to face. The recent Erasmus+ project *Pushing the boundaries of Online Transformative Learning* (OnTL) (2019–2020) will identify some of these specific challenges and explore the potential and limitations of virtual environments on empowering students in transformative action. Educators of adult learning and higher education from across the world (primarily Europe) will design and carry out a wide variety of experiments trying out different pedagogical approaches, assessment methods, apps, tools, etc. with different program settings and characteristics (short- and long-term courses, small and large groups, etc.) to critically reflect on how best we can support learners in the development of their sustainability competencies. This is an area which requires further attention, as there are practically no studies that have analyzed the implications of teaching and learning online in relation to sustainability learning and the development of sustainability competencies.

Concluding Remarks

Significant progress in ESD has been achieved over the last decades; however, the state of its integration is still disparate across different regions and education levels and between education institutions. In order to create transformative learning environments and experiences that facilitate the development of sustainability competencies, whole-institution approaches towards embedding sustainability are needed, embracing the estates and operations, the curriculum, pedagogy, the organizational structure and ethos.

Competence-based education entails moving from teacher-centered to student-centered approaches in combination with community and transformative learning processes that facilitate the development of sustainability competencies and lead to empowered and active change agents. In this context, the design and inclusion of ESD competencies within teacher education programs and through continuous professional development is critical to embed ESD processes and practices holistically through all education levels.

Several ESD scholars have made an effort to conceptualize sustainability competencies, skills, capacities, or learning outcomes; however, no agreed or validated framework exists that transcends education levels, examples of good practice, single case studies, or specific comparisons among universities. Further efforts are needed to develop common sustainability competencies' frameworks that can be tested and contextualized in different education levels and sociocultural settings. The operationalization of the term sustainability competencies remains as its main challenge. Further empirical research is critical to obtain evidence on innovative pedagogical and (trans)formative assessment approaches and strategies that lead to meaningful student learning and sustainability competencies' acquisition. Exploring how this learning takes place in other settings, such as non-formal, informal and community-based learning, and in professional contexts and the workplace, would provide a rich overview to make more informed pedagogical and curriculum decisions.

References

Ally, M. (2008). Foundations of educational theory for online learning. In T. Anderson (Ed.), *Theory and practice of online learning* (pp.15-44). Athabasca, AB: AU Press, Athabasca University.

Bertschy, F., Künzli, C., & Lehmann, M. (2013). Teachers' competencies for the implementation of educational offers in the field of education for sustainable development. *Sustainability, 5*(12), 5067–5080. https://doi.org/10.3390/su5125067

Black, P., & William, D. (1998). Assessment and classroom learning. *Assessment in Education: Principles, Policy and Practice*, *5*(1), 7–73.

Brundiers, K., Barth, M., Cebrián, G., Cohen, M., Diaz, L., Doucette-Remington, S., Dripps, W., Habron, G., Harré, N., Jarchow, M., Losch, K., Michel, J., Mochizuki, Y., Rieckmann, M., Parnell, R., Walker, P., & Zint, M. (2021). Key competencies in sustainability in higher education—toward an agreed-upon reference framework. *Sustainability Science, 16*, 13-29. https://doi.org/10.1007/s11625-020-00838-2

Bürgener, L., & Barth, M. (2018). Sustainability competencies in teacher education: Making teacher education count in everyday school practice. *Journal of Cleaner Production*, *174*, 821–826. https://doi.org/10.1016/j.jclepro.2017.10.263

Cebrián, G., & Junyent, M. (2015). Competencies in education for sustainable development: Exploring the student teachers' views. *Sustainability, 7*(3), 2768–2786. https://doi.org/10.3390/su7032768

Cebrián, G., Junyent, M., & Mulà, I. (2020). Competencies in education for sustainable development: Emerging teaching and research developments. *Sustainability, 12*(2), 579. https://doi.org/10.3390/su12020579

Dai, Y. & Hwang S. H. (2019). Technique, Creativity, and Sustainability of Bamboo Craft Courses: Teaching Educational Practices for Sustainable Development. *Sustainability, 11*(9), 2487. https://doi.org/10.3390/su11092487

de Haan, G. (2006). The BLK '21' programme in Germany: A 'Gestaltungskompetenz'-based model for Education for Sustainable Development. *Environmental Education Research*, *12*(1), 19–32. https://doi.org/10.1080/13504620500526362

Demssie, Y. N., Wesselink, R., Biemans, H. J. A., & Mulder, M. (2019). Think outside the European box: Identifying sustainability competencies for a base of the pyramid context. *Journal of Cleaner Production*, *221*, 828–838. https://doi.org/10.1016/j.jclepro.2019.02.255

Demssie, Y. N., Biemans, H. J. A., Wesselink, R., & Mulder, M. (2020). Combining indigenous knowledge and modern education to foster sustainability competencies: Towards a set of learning design principles. *Sustainability, 12*(17), 6823. https://doi.org/10.3390/SU12176823

Edwards, R. (2016). Competence-based education and the limitations of critique. International *Journal of Training Research, 14*(3), 244-255. https://doi.org/10.1080/14480220.2016.1254366

Glasser, H., & Hirsh, J. (2016). Toward the development of robust learning for sustainability core competencies. *Sustainability*, *9*(3), 178–184. https://doi.org/10.1089/SUS.2016.29054

Hohlfeld, T. N., Ritzhaupt, A. D., Dawson, K., & Wilson, M. L. (2017). An examination of seven years of technology integration in Florida schools: Through the lens of the Levels of Digital Divide in Schools. *Computers & Education, 113*, 135-161. https://doi.org/10.1016/j.compedu.2017.05.017

Holdsworth, S., Sandri, O., Thomas, I., Wong, P., Chester, A., & McLaughlin, P. (2020). The use of the theory of planned behaviour to assess graduate attributes for sustainability. *Environmental Education Research*, *26*(2), 275–295. https://doi.org/10.1080/13504622.2019.1700218

Leicht, A., Heiss, J., & Byun, W.J. (2018). *Issues and Trends in Education for Sustainable Development.* Paris, France: UNESCO. Available at: https://unesdoc.unesco.org/ark:/48223/pf0000261445

Lim, C. P., Zhao, Y., Tondeur, J., Chai, C., & Tsai, C. (2013). Bridging the gap: Technology trends and use of technology in schools. *Educational Technology & Society, 16*(2), 59–68. http://citeseerx.ist.psu.edu/viewdoc/download?doi=10.1.1.299.6895&rep=rep1&type=pdf#page=64

Lotz-Sisitka, Heyla. (2014). Can MOOCs change the world? *IAU Horizons*, *20*(1), 28–29. www.iau-aiu.net

Lotz-Sisitka, Heila, Wals, A. E. J., Kronlid, D., & McGarry, D. (2015). Transformative, transgressive social learning: Rethinking higher education pedagogy in times of systemic global dysfunction. *Current Opinion in Environmental Sustainability*, *16*, 73–80. https://doi.org/10.1016/j.cosust.2015.07.018

Lozano, J. F., Boni, A., Peris, J., & Hueso, A. (2012). Competencies in Higher Education: A Critical Analysis from the Capabilities Approach. *Journal of Philosophy of Education*, *46*(1), 132–147. https://doi.org/10.1111/j.1467-9752.2011.00839.x

Lozano, R., Merrill, M. Y., Sammalisto, K., Ceulemans, K., & Lozano, F. J. (2017). Connecting competences and pedagogical approaches for sustainable development in higher education: A literature review and framework proposal. *Sustainability, 9*(10), 1–15. https://doi.org/10.3390/su9101889

Lozano, R., Barreiro-Gen, M., Lozano, F. J., & Sammalisto, K. (2019). Teaching sustainability in European higher education institutions: Assessing the connections between competences and pedagogical

approaches. *Sustainability, 11*(6), 1–17. https://doi.org/10.3390/su11061602
McLaren, P. (1994). Foreward: Critical thinking as a political project. In K. Walters (Ed.), *Re- thinking reason: new perspective in critical thinking* (pp. ix–xv). New York: State University of New York Press.
Mindt, L., & Rieckmann, M. (2017). Desarrollo de las competencias para el emprendimiento orientado a la sostenibilidad en la educación superior: una revisión bibliográfica de los métodos de enseñanza y aprendizaje. *Teoría de la Educación. Revista Interuniversitaria, 29*(1), 129-159. https://doi.org/10.14201/teoredu291129159
Mogensen, F. and Schack, K. (2010). The action competence approach and the 'new' discourses of education for sustainable development, competence and quality criteria. *Environmental Education Research, 16*(1), 59-74. https://doi.org/10.1080/13504620903504032
O'Donoghue, R., Lotz-Sisitka, H., Asafo-Adjei, R., Kota, L., & Hanisi, N. (2007). Exploring learning interactions arising in school-in community contexts of socio-ecological risk. In A. E. Wals (Ed.), *Social learning towards a sustainable world* (pp. 435–449). Wageningen: Wageningen Academic Publishers.
O'Sullivan, E. (2003). Bringing a perspective of transformative learning to globalized consumption. *International Journal of Consumer Studies, 27*(4), 326-330. https://doi.org/10.1046/j.1470-6431.2003.00327.x
O'Sullivan, E., Morrell, A. & O'Connor, M.A. (Eds.). *Expanding the boundaries of transformative learning: Essays on theory and praxis.* New York: Palgrave Macmillan. 2002
Panigrahi, R., Srivastava, P. R., & Sharma, D. (2018). Online Learning: Adoption, Continuance, and Learning Outcome—A Review of Literature. *International Journal of Information Management, 43*, 1-14. https://doi.org/10.1016/j.ijinfomgt.2018.05.005
Popham, W. J. (2008). *Transformative assessment.* Alexandria, VA: Association for Supervision & Curriculum Development (CSCD).
Rauch, F., & Steiner, R. (2013). Competences for Education for Sustainable Development in Teacher Education. *CEPS Journal - Center for Educational Policy Studies Journal, 3*(1), 9–24.
Redman, A., Wiek, A. & Barth, M. (2021). Current practice of assessing students' sustainability competencies: a review of tools. *Sustainability Science 16*, 117-135. https://doi.org/10.1007/s11625-020-00855-1
Rieckmann, M. (2012). Future-oriented higher education: Which key competencies should be fostered through university teaching and learning? *Futures, 44*(2), 127–135. https://doi.org/10.1016/j.futures.2011.09.005
Rieckmann, M. (2013). The global perspective of education for sustainable development: a European-Latin American study about key competencies for thinking and acting in the world society. *Environmental Education Research, 19* (2): 257-258. https://doi.org/10.1080/13504622.2012.697547
Ryan, A., & Tilbury, D. (2013). Uncharted waters: Voyages for Education for Sustainable Development in the higher education curriculum. *Curriculum Journal, 24*(2), 272–294. https://doi.org/10.1080/09585176.2013.779287
Sadler, D. R. (2013). Making competent judgments of competence. In S. Blömeke, O. Zlatkin-Troitschanskaia, C. Kuhn & J. Fege (Eds.), *Modeling and measuring competencies in higher education: Tasks and challenges* (pp. 13–27). Rotterdam: Sense Publishers.
Sen, A. (1993). Capability and Wellbeing. In M. Nussbaum & A. Sen (Eds.), *The quality of life* (pp. 30–55). Oxford: Clarendon Press.
Shephard, K., Rieckmann, M., & Barth, M. (2019). Seeking sustainability competence and capability in the ESD and HESD literature: an international philosophical hermeneutic analysis. *Environmental Education Research, 25*(4), 532–547. https://doi.org/10.1080/13504622.2018.1490947
Sleurs, W. (2008). *Competencies for ESD (Education for Sustainable Development) teachers: A framework to integrate ESD in the curriculum of teacher training institutes - Comenius 2.1 project 118277-CP-1-2004-BE-Comenius-C2.1.* Available at: http://www.unece.org/fileadmin/DAM/env/esd/inf.meeting.docs/EGonInd/8mtg/CSCT%20Handbook_Extract.pdf
Sterling, S. (2001). *Sustainable Education: Re-Visioning Learning and Change. Schumacher Society Briefing no 6.* Dartington: Green Books.
Sterling, S., Glasser, H., Rieckmann, M., & Warwick, P. (2017). "More than scaling up": a critical and practical inquiry into operationalizing sustainability competencies. *Envisioning Futures for Environmental and Sustainability Education, August*, 153–168. https://doi.org/10.3920/978-90-8686-846-9_10
Strachan, G. (2012). *WWF professional development framework of teacher competences for learning for Sustainability.* London: WWF-UK.
Thomas, I. (2009). Critical Thinking, Transformative Learning, Sustainable Education, and Problem-Based Learning in Universities. *Journal of Transformative Education, 7*(3), 245-264. https://doi.org/10.1177/1541344610385753
Tilbury, D. (2011), *Education for Sustainable Development: An Expert Review of Processes and Learning,* UNESCO, Paris. Available at: http://unesdoc.unesco.org/images/0019/001914/191442e.pdf
Tilbury, D.; Mulà, I. (2009). *Review of Education for Sustainable Development Policies from a Cultural Diversity and Intercultural Dialogue: Gaps and Opportunities for Future Action.* Paris: UNESCO.
Timm, J. M., & Barth, M. (2021). Making education for sustainable development happen in elementary schools: the role of teachers. *Environmental Education Research, 27*(1), 50-66. https://doi.org/10.1080/13504622.2020.1813256
UN. (2015). *Transforming our workd: The 2030 Agenda for sustainable development. Resolution Adopted by the General Assembly on 25 September 2015 (A/*

RES/70/L.1). Available at: http://sustainabledevelopment.un.org/post2015/transformingourworld

UN. (2019). *Education for sustainable development in the framework of the 2030 Agenda for Sustainable Development.* United Nations. Available at: https://undocs.org/en/A/RES/74/223

UNECE. (2012), *Learning for the future: Competences in Education for Sustainable Development,* UNECE, Geneva. Available at: http://www.unece.org/fileadmin/DAM/env/esd/ESD_Publications/Competences_Publication.pdf

UNESCO. (2004), *United Nations Decade of Education for Sustainable Development (2005-2014): Draft International Implementation Scheme*, UNESCO, Paris. Available at: http://portal.unesco.org/education/en/file_download.php/e13265d9b948898339314b001d91fd01draftFinal+IIS.pdf

UNESCO. (2014). *Shaping the future we want. DESD monitoring and evaluation UN Decade of Education for Sustainable Development (2005-2014). Final report,* UNESCO, Paris. Available at: https://sustainabledevelopment.un.org/index.php?page=view&type=400&nr=1682&menu=35

UNESCO. (2015). *Rethinking education : towards a global common good?*, UNESCO, Paris. Available at: https://unesdoc.unesco.org/ark:/48223/pf0000232555

UNESCO (2017), *Education for Sustainable Development Goals. Learning Objectives*, UNESCO, Paris. Available at: http://unesdoc.unesco.org/images/0024/002474/247444e.pdf

UNESCO. (2020). *Education for sustainable development. A roadmap*, UNESCO, Paris. Available at: DOI: https://doi.org/10.1111/j.2048-416x.2009.tb00140.x

Vare, P., Arro, G., de Hamer, A., Gobbo, G. Del, de Vries, G., Farioli, F., Kadji-Beltran, C., Kangur, M., Mayer, M., Millican, R., Nijdam, C., Réti, M., & Zachariou, A. (2019). Devising a competence-based training program for educators of sustainable development: Lessons learned. *Sustainability, 11*(7) 1890. https://doi.org/10.3390/su11071890

Wals, A. E. J. (Ed.) (2007). *Social Learning towards a Sustainable World.* Wageningen: Wageningen Academic Publishers.

Wals, A. E. J. (2020). Transgressing the hidden curriculum of unsustainability: towards a relational pedagogy of hope. *Educational Philosophy and Theory*, *52*(8), 825–826. https://doi.org/10.1080/00131857.2019.1676490

Wiek, A., Withycombe, L., & Redman, C. L. (2011). Key competencies in sustainability: A reference framework for academic program development. *Sustainability Science*, *6*(2), 203–218. https://doi.org/10.1007/s11625-011-0132-6

Wiek, A., Xiong, A., Brundiers, K. & van der Leeuw, S. (2014). Integrating problem- and project-based learning into sustainability programs: A case study on the School of Sustainability at Arizona State University. *International Journal of Sustainability in Higher Education, 15*(4), 431-449. https://doi.org/10.1108/IJSHE-02-2013-0013

Wiek, A., Bernstein, M.J., Foley, R.W.; Cohen, M.; Forrest, N., Kuzdas, C. et al. (2016): Operationalising competencies in higher education for sustainable development. In M. Barth, G. Michelsen, I. Thomas & M. Rieckmann (Eds.), *Routledge Handbook of Higher Education for Sustainable Development* (pp. 241-260). London: Routledge.

Wilhelm, S., Förster, R., & Zimmermann, A. B. (2019). Implementing competence orientation: Towards constructively aligned education for sustainable development in university-level teaching-and-learning. *Sustainability, 11*(7), 1891. https://doi.org/10.3390/su11071891

Yoneyama, S. (2012). Critiquing critical thinking: Asia's contribution towards sociological conceptualisation. In X. Song & K. Cadman (Eds.), *Bridging transcultural divides: Asian languages and cultures in global higher education* (pp. 231–252). Adelaide: University of Adelaide Press.

Zhang, W., Wang, Y., Yang, L., & Wang, C. (2020). Suspending classes without stopping learning: China's education emergency management policy in the COVID-19 outbreak. *Journal of Risk and Financial Management*, 13(3), 55. https://doi.org/10.3390/jrfm13030055

23 Closing Thoughts: The Role of Educator Competences for ESD in Turbulent Times

Paul Vare, Nadia Lausselet, and Marco Rieckmann

Abstract

This chapter takes stock of some of the lessons of the COVID-19 pandemic and relates these back to the core messages of this book. One 'take away' from this discussion is the possibility of rapid change, which echoes what is needed for education for sustainable development (ESD), another is the fact that people frequently claim to have learned things from the pandemic that reinforce their pre-existing beliefs, showing a tension between evolution and conservation. The chapter closes by identifying three inter-related challenges confronting the effort to integrate ESD into mainstream formal education: firstly, educators need competences that extend beyond the attributes required of those in mainstream settings today; secondly there will need to be a shift in priorities across institutions to support this. Finally, we need to achieve the political will, at the level of education policymakers, to enshrine *contributing to the imperative of securing a socially just and ecologically sustainable future* within the core purposes of education.

Keywords

Competences · COVID 19 · Education for sustainable development · Purpose of education

Learning from COVID?

The fact that we have been planning and writing this volume during 2020 and 2021 is not an insignificant detail. Over this time, we have all been experiencing periods of lockdown due to the latest pandemic to afflict humanity, a coronavirus, identified in 2019, known to us now as COVID-19.

As well as causing the tragic loss of life and livelihoods, the COVID-19 pandemic has disrupted the education of millions of children and young people worldwide. In the wake of this severe loss of (formal) learning, the UN has launched a policy brief that calls for disease suppression, financial support and plans to *strengthen the resilience of education systems for equitable and sustainable development* (UNESCO – United Nations Educational, Scientific and Cultural

P. Vare (✉)
University of Gloucestershire, Cheltenham, UK
e-mail: pvare@glos.ac.uk

N. Lausselet
University of Teacher Education, Lausanne, Vaud, Switzerland

M. Rieckmann
University of Vechta, Vechta, Germany

P. Vare et al. (eds.), *Competences in Education for Sustainable Development*, Sustainable Development Goals Series, https://doi.org/10.1007/978-3-030-91055-6_23

Organisation 2021, p. 3). This might be seen as a promising outcome but there have been international calls for education to take sustainability seriously since at least 1969.[1] What this call means for educators, beyond wider access to and better management of education, is not entirely clear—a point we return to below.

The well-being of children and young people should be a high, if not our top, priority. Unfortunately, this has not only failed to be the case in our response to the pandemic but more generally, the underfunding of education remains a stubborn reality across the globe. Even in relatively wealthy Western countries, we find classrooms that are small and poorly ventilated, technology that has proved to be outdated and unable to cope and teachers that could benefit from strengthened professional development. That said, significant steps have been taken with better funding for digital equipment in many schools while, accelerated by the pandemic, the upgrading of digital technologies and related learning in the higher education system has occurred at an unimaginable pace (Leal Filho et al. 2021). Digitalisation however, does not equal well-being of children or quality education; furthermore, the inconsistent distribution of educational provision throughout the pandemic is likely to increase inequalities (Cecchini and Dutrévis 2020).

Digitalisation in education does not equate to climate neutrality either. It certainly has the potential to help reduce ecological impacts by avoiding travelling for example but within the field of education for sustainable development (ESD), all this needs to be looked at critically so as to ensure that it contributes to a quality education in coherence with sustainability concerns.

In terms of what the pandemic has taught us in relation to learning more broadly, the possibility of rapid change must surely be a key lesson. As Kaukko et al. (2021) put it, 'Worldwide responses to the Coronavirus have demonstrated that vast upheavals of existing arrangements are feasible, after all' (*Ibid*, p. 12). This shows that it is entirely possible to respond to scientific-based projections in the short term with political and social action. Longer term, the nature and climate crises threaten even more serious consequences for human health than COVID-19, particularly in poorer countries with one-third of the world's population likely to be experiencing mean annual temperatures greater than 29 °C within the next 50 years, a situation currently found in only 0.8% of the globe (Xu et al. 2020). Little wonder that many hope that when the worst effects of the pandemic have past, evidence-based warnings, such as those of climate scientists will be taken equally seriously, 'not least because deforestation, pollution, biodiversity loss are all contributory factors to the spread of the virus' (UN – United Nations 2020, p. 23).

According to Beasley and Gonzalez (2021), who examined the perceptions of their education community in Australia, reactions to seeing rapid responses to COVID-19 have been broadly positive and optimistic. Reaction may have been swift but not all authors see this so positively. Among them, Duflot et al. (2021) claim that the pandemic has been a 'crash test' for our resilience in an uncertain future and that this has revealed how poorly prepared humanity is to cope with such stress tests. The political establishment's call to *build back better* can mask an eagerness to return to 'business as usual' as quickly as possible. The UK Government's plan for a 'Green Industrial Revolution', for example, proposes spending a hundred billion pounds on infrastructure against only 12 billion pounds for projects to achieve 'net zero' (UK Govt 2021).

A further lesson we might take away from these different responses to the pandemic is that people may claim to have learned things that reinforce their pre-existing beliefs. Rapid change has certainly taken place but we cannot assume that this will continue or necessarily be in a positive direction. Indeed, Lehmann et al. (2021) warn us that, 'a naive opportunity narrative may even impair the progress of transitions towards environmental sustainability' (*Ibid*, p. 2). We cannot afford to relax in our efforts to transform education so that it can, in turn, empower the next generation of learners to contribute fully to the

[1] For a full account of these calls, leading up to the Tbilisi Declaration see Chapter 25 in Scott and Vare (2021).

transformation of society towards a more just and ecologically sustainable model of development.

The Role of Educator Competences for ESD

Education cannot, of course, be expected to solve the world's ills, hence we are careful to use the term 'contribute'. In light of the foregoing discussion however, we are more convinced than ever of the need for a reorientation of formal education that will *inter alia* involve a recasting of the role of the educator and the competences that are key to that new role. In Part I of this book, we have discussed advantages but also limitations of the competence-based approach, both in a general sense and in its specific application to ESD. Part II has explored the double-edged nature of competence frameworks—as a means of structuring effective staff development programmes but also as an imposition that can arouse suspicion and resistance. In Part III we have focused on pedagogical approaches and discussed how these need to be applied thoughtfully to support the development of ESD competences although in the available space we could only introduce a small sample of approaches. There is so much more to explore, from subject-specific innovations such as in Mathematics (Coles 2016) or history (Hendriawan et al. 2019) to inter- and trans-disciplinary approaches such as student-led school projects and locally relevant teaching (Vare 2021; Van Poeck and Östman 2020).

The impact of the pandemic will have influenced these practices just as it has impacted the work of the contributors to this book. Our programmes of study have been digitised, provided remotely and, in some cases, offered in 'blended' combinations of face-to-face and online learning. In some cases, courses have taken place entirely outdoors and we have witnessed something of a boom in outdoor education in some countries such as Switzerland and France. Combined modes of teaching have been investigated before the pandemic of course but there has been a massive scaling up and it seems much of this 'new' practice may well remain, including in some surprising areas such as outdoor education (OE) where one 2017 study revealed how:

> ...with careful design and delivery, the online space (accompanied by a single field experience for some learners) was an effective way to deliver a foundational OE unit (Dyment et al. 2018, p. 81).

This is a promising finding in its own right for those, who for whatever reason, wish to provide outdoor learning under constrained circumstances, showing that ESD and digitalisation can enjoy a fruitful partnership. However, for any new practice to be adopted and sustained, it will need to be truly embedded within a supportive system. This is particularly true of educator competences, the development of which extends well beyond the scope of course timetables. Just as institutions have had to adapt to new circumstances imposed by the pandemic, so they will need to consider carefully the implications that teaching, learning and assessing ESD competences will have for their structures and practices. As many of the case studies in this book demonstrate, this is a gradual process rather than a rapid imposition and it relies as much on bottom-up pressures from educators themselves as it does on supportive political contexts to be effective. Such whole institutional approaches go beyond thinking holistically in relation to a specific programme of study; they include working on and within the institutional environment, which in turn requires a parallel process of organisational learning.

Three Levels of Challenge

In this book we have explored the case for developing educators' competences in ESD and shared in the practice of doing so in a range of contexts across Europe; yet the scale at which this needs to happen for education to support the transition to a more sustainable society is daunting. The challenge lies at three levels:

Firstly, educators will need competences that extend beyond the attributes required of those in most mainstream settings today; the challenge is to identify the *what* and the *how* of doing this in each context. This will include their perception of themselves as change agents. We trust that this volume will inform this debate.

The second challenge lies at the institutional level. Shifting the role of educators—and thus education—will demand time, resources and institutional flexibility. It also demands seeing the institution itself as a learning environment, thus creating a supportive ecology of learning (Stratford & Wals 2020) for developing ESD competences. Justifying all this in the face of short-term demands and standardised measures of accountability will require a shift in priorities more generally across most institutions. While not providing details on how this can best be achieved, this volume does, we hope, provide a rationale for engaging in such a process.

Addressing the second challenge would be facilitated to a significant extent by progress on the third, that is, achieving the political will, at the level of education policymakers, to enshrine *contributing to the imperative of securing a socially just and ecologically sustainable future* within the core purposes of education. This is no small task but our efforts in relation to the other two challenges can remain piecemeal unless this third challenge is taken seriously. To achieve this for the long term will require broad agreement across ideological boundaries and we are some way off that point. In too many contexts, the debates about the purpose of education are silenced by a taken-for-granted assumption that education operates as an arm of the economy.[2] Pressure from beneath is important here, just as gaining new competences can transform educators, so the process can embolden educators to become change agents who encourage their institutions to become a part of the process. Nobody need feel it is all up to them, neither should they be waiting for others to take the lead. Taking part is critical, for if this situation is not addressed in the near future, it may be too late for ESD; indeed, there are already calls for reorienting education for 'the end of the world as we know it' (Stein et al. 2020).

The existential risk involved in allowing our education systems to simply recreate our profoundly unsustainable model of development places a responsibility on all of us to do what we can to tackle these challenges. Competences are no silver bullet but they do provide a useful means of linking a concern for social justice and the ecological integrity of our planet with the everyday practices of educators and ultimately the lessons they provide for their learners. This book can inform actors across education from policymakers to curriculum developers, university administrations, teacher educators and teachers, about what might be done. Our hope is that our practical efforts will be supported by political demands to secure an education with a rounder sense of purpose.

[2] A current example of this can be found in the UK where the Government's *Projected Completion and Employment from Entrant Data* (Proceed) records the nature of jobs (and income) secured by higher education alumni and uses this as a key measure of the 'quality' of education offered by each institution.

References

Beasley, K. & Gonzalez, L.R. (2021). Exploring Changes in Perceptions and Practices of Sustainability in ESD Communities in Australia during the COVID-19 Pandemic. *Journal of Education for Sustainable Development,* 15(1):5-24. https://doi.org/10.1177/09734082211012081

Cecchini, A. & Dutrévis, M. (2020). Le Baromètre de l'école – Enquête sur l'école à la maison durant la crise sanitaire du Covid-19. https://www.ge.ch/document/21573/telecharger (10.09.21)

Coles, A. (2016) Mathematics education in the Anthropocene. *Proceedings of the third conference of mathematics education and contemporary theory.* http://www.esri.mmu.ac.uk/mect3/papers_16/coles.pdf

Duflot, R., Baumeister, S., Burgas, D., Eyvindson, K., Triviño, M., Blattert, C. Kuparinen, A. & Potterf, M. (2021). Building up an ecologically sustainable and socially desirable post-COVID-19 future. *Sustain Sci* 16, 1397–1403, https://doi.org/10.1007/s11625-021-00940-z

Dyment, J., Downing, J., Hill, A. & Smith, H. (2018). 'I did think it was a bit strange taking outdoor education online': exploration of initial teacher education students' online learning experiences in a tertiary outdoor education unit, *Journal of Adventure Education and Outdoor Learning*, 18:1, 70-85, https://doi.org/10.1080/14729679.2017.1341327

Hendriawan, D., Ali, M. & Rusman (2019). High School History Education and Education for Sustainable

Development. An Integrated Curriculum Approach J. *Phys.: Conf. Ser.* 1179 012048

Kaukko, M., Kemmis, S., Heikkinen, H.L.T., Kiilakoski, T. & Haswell N (2021). Learning to survive amidst nested crises: can the coronavirus pandemic help us change educational practices to prepare for the impending eco-crisis? *Env. Educ. Res.*, https://doi.org/10.1080/13504622.2021.1962809

Leal Filho, W., Price, E., Wall, T. *et al.* (2021). COVID-19: the impact of a global crisis on sustainable development teaching. *Environ Dev Sustain* 23, 11257–11278 https://doi.org/10.1007/s10668-020-01107-z

Lehmann, P., de Brito, M.M., Gawel, E. *et al.* (2021). Making the COVID-19 crisis a real opportunity for environmental sustainability. Sustain Sci. https://doi.org/10.1007/s11625-021-01003-z

Scott W.A.H. & Vare P. (2021) *Learning, Environment and Sustainable Development: a history of ideas.* Abingdon: Routledge. https://www.routledge.com/Learning-Environment-and-Sustainable-Development-A-History-of-Ideas/Scott-Vare/p/book/9780367221935

Stein, S., Andreotti, V., Suša, R., Ahenakew, C. & Čajková, T. (2020). From "education for sustainable development" to "education for the end of the world as we know it." *Educational Philosophy and Theory*. https://doi.org/10.1080/00131857.2020.1835646

Stratford, R. & Wals, A. E. (2020). In search of healthy policy ecologies for education in relation to sustainability: Beyond evidence-based policy and post-truth politics. *Policy Futures in Education, 0(0) 1–19.* https://doi.org/10.1177/1478210320906656

UK Govt (2021). *Build Back Better: our plan for growth.* London: HM Govt.uk. https://www.gov.uk/government/publications/build-back-better-our-plan-for-growth

UN – United Nations (2020) *Shared Responsibility, Global Solidarity: Responding to the socio-economic impacts of COVID-19.* New York: UN. https://unsdg.un.org/sites/default/files/2020-03/SG-Report-Socio-Economic-Impact-of-Covid19.pdf

UNESCO – United Nations Educational, Scientific and Cultural Organisation (2021). *Policy Brief: Education during COVID-19 and beyond.* Paris: UNESCO https://www.un.org/sites/un2.un.org/files/sg_policy_brief_covid-19_and_education_august_2020.pdf

Van Poeck, K. & Östman, L. (2020). The Risk and Potentiality of Engaging with Sustainability Problems in Education—A Pragmatist Teaching Approach. Journal of Philosophy of Education, Vol. 54, No. 4. https://doi.org/10.1111/1467-9752.12467

Vare, P. (2021). Exploring the Impacts of Student-Led Sustainability Projects with Secondary School Students and Teachers. *Sustainability*, 13, 2790. https://doi.org/10.3390/su13052790

Xu, C., Kohler, T.A., Lenton, T.M., Svenning, J-C. & Scheffer, M. (2020). "Future of the Human Climate Niche." *Proceedings of the National Academy of Sciences* 117 (21): 11350–11355. https://www.pnas.org/content/117/21/11350

Index

A
Abilities, 46
Able, 51
Academic staff training
German higher education, 122–123
Accountability, 25
Achieving transformation, 23
Action-oriented and transformative approach, 21
Action-oriented approaches, 146
Action research, 89, 90
Active non-activeness, 156
Aesthetic learning, 146, 147
Agency, 155
Agents of change, 63
AISHE, 106, 110
A Rounder Sense of Purpose (RSP), 6, 22
Artificial snowmaking system, 56
Arts-based sustainability education
approaches, 148
challenges and needs, 155, 156
Assessment, 94, 96, 98
Attentive listening, 154
Attentiveness, 36, 95
Attitudes, 48

B
Behaviours, 48
Being willing, 49
Bildung, 4

C
Capability, 187
Centrality of context, 14, 15
Change agent, 24, 86, 87, 89
Circumstantial paradigm, 176
Citizenship education, 53
Classroom, 21
Collaborative competencies, 154
Collective intelligence, 171
Community, CBL, 110
Competence, 46, 48
and competency, 4
concepts of, 4
coverage, 28
definition, 4, 5
frameworks, 148, 152
Competence-based approach, 57, 58, 84
Competence-based education, 12, 72, 191
Competence-based education for sustainable development (ESD), 121
Competence-based learning (CBL), 14, 104
ESD, analysis of implementation, 106, 109–111
community, 110
facility management, 109
Human Resource Management, 109
professional development, 109, 110
teaching, 109
Marnix Academie (MA) and, 106
methodology, 104
Netherlands, Teacher Training Institutes in, 105, 106
Competences, 121–127, 187
ability/trait, 5
Bildung, 4
concept of, 4
defining, 4
ESD (*see* Education for sustainable development (ESD))
European Union, 6
framework, 15
language, 5
learning outcome, 4
literacy, 6
literature review, 6
SD, 140–142
shaping competence, 5
transformation, 5
Competence Turn, 7
centrality of context, 14, 15
competence-based education, 12
conceptual confusion, 14
curriculum, 11
Digicomp, 11

P. Vare et al. (eds.), *Competences in Education for Sustainable Development*, Sustainable Development Goals Series, https://doi.org/10.1007/978-3-030-91055-6

Competence Turn (*cont.*)
distractions from fundamental issues, 16, 17
Entrecomp, 11
ESE, 12, 13
frameworks, 11
impacts on pedagogy, 15, 16
LifeComp, 11
Competencies, 4, 46, 177, 187
linked to, 152
sustainability
in agreement, 27, 28
current practices, flaws in, 28–31
good practices, 31, 32
technical meaning, 47, 48
Competency-based education and training (CBET), 12, 47
Complex thinking, 55
at school, 55
Computer-RPGs (CRPGs), 162
Concept mapping, 4–6, 55
Conceptual awareness, 13
Conceptual confusion, 14
Conspiracy theory, 58, 59
Content and teaching and learning methods, 20
Content knowledge (CK), 23, 46
Controversial issues (CI), 53
Cooperation, 189
Crash test, 196
Criticality, 36, 95
Critical review, 19, 23
Critical thinking, 23, 55, 154
acting, 140
competencies, 154
at school, 56, 57
skills, 51
Critique of competences, 14
CSCT model, 21–23
Curriculum approaches, 11–13, 15, 16
Curriculum as content, 12, 16
Curriculum as process, 15
Curriculum as product, 12, 15, 16
Curriculum, Sustainable development, Competences, Teacher (CSCT) training, 13
Cyprus
educational context, 134
professional development and ESD in, 130
professional development in, 134

D
Decisiveness, 38
Deep quality in education
RSP model of ESD competences and, 73, 74
Definition and Selection of Competencies (DeSeCo), 13
Digicomp, 11
Digital communities of learning, 133
Digital education, 133
Dispositions, 4, 48, 49, 51

E
Ecological foundations, 13
Educating Future Change Agents (EFCA), 176, 181
Educational Authority of Hungary, 95
Educational games, 160
Educational mini-games, 162
Educational praxis, 24
Educational programmes, 20
Education for sustainability (EfS), 85
Education for sustainable development (ESD), 36, 45, 51, 61, 66, 104, 121, 129, 140, 159, 176–178, 186
blunted conceptual tool, in the hands of, 49
CBL, 110, 111
competences, 131
and methodology, 131
educator competences for
challenge, three levels of, 197, 198
COVID, 195–197
role of, 197
educators' professional development, impact on, 132
ESE, 4
factor, 132
familiarity with the RSP model, 132
Gestaltungskompetenz, 5
HEP Vaud, action plan for, 117
integration of, 127
and LfS, 4
paradigm shift, 114
pre-service primary school teachers, course for, 114–117
professional development and, 130, 131
SE, 4
supportive context for, 113, 114
teacher, PD on, 130
training received, 131, 132
UOC's competence framework and, 80–82
Education for the Environment and Sustainable Development (EESD) Unit, 130
Educator competences, 3, 4, 6–8, 94, 95
challenge, three levels of, 197, 198
COVID, 195–197
frameworks and models, 21–23
role of, 197
Educators, 19, 85
professional development, ESD, 132
promoting change, 86
RSP competence model, 88, 89
sustainability education, for Italian context, 86, 87
Emotions management, 23
Empathy, 5
Entrecomp, 11
Environmental action skills, 13
Environmental and sustainability education (ESE), 4, 12–14
Environmental autobiography, 89
Environment and School Initiatives (ENSI), 13
Equipping educators, 21

Error reproduction, 30
ESD competences
  competence-based approach, 57, 58
  complexity and criticality
    citizenship education, competences, 53
    paradigms of, 54, 55
  complex thinking at school, 55
  conspiracy, 58, 59
  critical thinking at school, 56, 57
  development and assessment of, 123–124
  frameworks
    analysis, 24
    elements, 21
    and models, 23
  highs and lows of, 80
    tensions and opportunities, 82, 83
  limitations, 126
  QE, 69, 70
    competences and, 71–73
    economic model, 70
    humanist paradigm, 70, 71
    learning as connection, 71
    RSP model, 73, 74
ESD educator competence
  mainstreaming an, 93
  pedagogical questions, 95
  structural questions, 99, 100
ESD educators, 8
ESD teachers, 22
Estrangement, 154
Ethical competence, 81, 82
European HEIs, 142
European Higher Education, 188
Experiential learning, 146, 147

F
Foresighted thinking, 5
Frame of mind, 16

G
Game design, 162, 163
Game Master, 163
Game Master designs, 162
Game-for-Learning, 160
Gamification, 159, 160
  approaches to, 160, 161
  competence, 160
  ethical considerations, 161
  game design, 162, 163
  phenomenon of, 160
Gamified learning environments, 161
Gaming sessions for learning, 162
Gender, 82, 83
German higher education
  academic staff training in, 122–123
  theoretical and practical approaches, 126
Gestaltungskompetenz, 5, 13
Green Industrial Revolution, 196
Green lobby, 16
Green sustainable development, 109

H
HEP Vaud, action plan for, 117
Higher education, 28
Higher education in particular (HESD), 45
Higher Education Institutions (HEIs), 79
Human Resource Management (HRM), 109

I
Impacts on pedagogy, 15, 16
Inclusive pluralism, 16
INDIRE, 90
Individual and group agency, 154
Individual transformation, 118
Institutional transformation, 118
Integration, 21, 28
Intelligence Quotient (IQ), 176
Interconnectedness, 152
Inter-governmental Conference on Environmental Education, 12
International declarations, 20
International Environmental Education Programme (IEEP), 13
International Standard Classification of Education (ISCED), 39
Interpersonal competencies, 154
Investigation and evaluation, 13
Italian Association for Sustainability Science (IASS), 62
  dealing with uncertainty, 64
  different, and circumstances and contexts, 63, 64
  PNS, making to educators, 63
  role of educators, 64
  transformation, role of educators, 65
  uncomfortable knowledge, 65
Italian National System on Environmental Information, Training and Education (INFEA), 86

K
Knowing competences, 133
KOM-BiNE model, 13, 21–23

L
Language competence, 5
Learner-centred approaches, 146
Learning for sustainability (LfS), 4
Learning outcomes, 4
  Rounder Sense of Purpose (RSP), 41
Legitimate aim, 16
LifeComp, 11
Likelihood of transformation, 177
Literacy, 6

M
Mainstreaming, 93
Marnix Academie (MA), 104, 106
Measurability, 25
Measurement, 176, 181
Millennium Development Goals (MDGs), 107
Ministry of the Environment, 88
Model of sustainability, 113

N
National policy papers, 20
National qualification competence framework, 96–99
National System for Environmental Protection (SNPA), 88, 90
Netherlands, Teacher Training Institutes in, 105, 106
Neurobiology, 161
Non-formal educators, 86, 87
Normative competency, 154

O
Objective-based learning, 16
Observation-based procedures, 32
One-size-fits-all approach, 32
Online learning, 190, 191
Online training, 81, 83, 84
Organisation for Economic Cooperation and Development (OECD), 13
Outdoor education (OE), 8, 167, 168
  analysis of students' point of view and transfer activity, 170
    collective intelligence, 171
    foster, active encounter, 170, 171
    outdoor experience and school knowledge, 171
  operationalizing transformative, 169, 170
  transformative outdoor education, 168, 169
Outdoor project-based approach, 169, 170

P
Paradigm revolution, 139
Paradigm shift, 114
Participation, 38, 189
Participatory art, 148
Pedagogical approach, 142, 143
  Rounder Sense of Purpose (RSP), 40
Pedagogical content knowledge (PCK), 23, 46, 181
Pedagogical model for a national system, 94
Pedagogies, 140
  slow, embodied, aesthetic, 146
Personal dispositions, 51
Post-Normal Science (PNS), 62–64, 66
Power of agency, 177
Pre-service primary school teachers, course for, 114–117
Prise, 106, 110
Professional action competence for teachers, 22
Professional context, 24
Professional development
  CBL, 109, 110
  and ESD, 130, 131
Professional discomfort, 87
Professional learning activities, 132
Program-level learning objectives (PLLOs), 30
Programme for International Student Assessment (PISA), 47, 180
Project-based co-constructed learning process, 171
Purpose of education, 12, 16, 198

Q
Qualifying educators, 21
Quality Education (QE), 69, 70
  economic model, 70
  ESD, competences and, 71–73
  humanist paradigm, 70, 71
  learning as connection, 71
  RSP model, 73, 74

R
Relational dimension, 152
Respective competence profiles, 22
Restricted ability-based conception of competence, 47
Role-playing games (RPGs), 162
Rounder Sense of Purpose (RSP), 15, 35, 55, 61, 62, 80, 90, 96–99, 113, 121, 130, 134, 170
  achieving transformation, 38
  competences, 20, 90
    educators, 88, 89
  envisioning change, 36, 38
  ESD, 73, 74, 132
  framework, 24, 36
  key issues/challenges, 39
    assessment, 41
    competence, 39, 40
    graphic representation, 40
    learning outcomes, 41
    pedagogical approach, 40
  Sustainable Development Goals (SDGs)
    as a means, 38, 39
    quality education, 38
  thinking holistically, 36

S
School knowledge, 171
Scopus, 147
Second order change, 86
Self-awareness, 152
Self-perceiving-based procedures, 32
Self-respecting educator, 17
Semantics, 49, 50
Sensorial imagination, 154
Sensory fieldwork, 168
Sensuous, embodied, critical pedagogies, 147

Serious games, 160
Shaping competence, 5
Skills/skill set, 4
Socio-ecosystems, 156
Staff training, 50
 at University of Vechta, 125–126
Startbekwaamheidseisen, 105
Status quo, 185
Storytracks, 160
Structure of knowledge, 133
Sustainability, 93, 187
 commitment, 16
 competencies in
  agreement, 27, 28
  current practices, flaws in, 28–31
  good practices, 31, 32
 education for
  Italian context for, 86, 87
 gamification for, 159, 160
  approaches to, 160, 161
  ethical considerations, 161
  game design, 162, 163
Sustainability competences, 19, 175–179
 assessment tool, 180
 challenges, 180
 EFCA project, 181
 interwoven assessment tools, 179
 in vivo simulation, 181
 outcomes, 180
 range of evidence, 180
 typology of tools, 178
Sustainability competencies, 148
 conceptual terminology and rich descriptions, 187
 context in, 189, 190
 current frameworks of, 154
 curriculum design and development, 188, 189
 pedagogical transformation and stimulate fundamental, 186
 research and experimentation, 190, 191
 status quo, 185
Sustainability education (SE), 4, 145, 167, 168
 artistic turn in, 146, 147
 explicit link to, 171, 172
 methodological approach, 147, 148
Sustainable development (SD), 19, 24, 113
 concept of, 3
 and role of education, 3–6
 state of the art of, 140–142
Sustainable Development Goals (SDGs), 3, 20, 36, 64, 71, 73, 80, 107
Sustainable education, 87
Systems thinking, 36, 189

T
Teacher education, 20
 ESD competences in, 119
Teacher–pupil interaction, 58
Teachers' and teacher educators, 129
Teacher trainers
 for education for sustainable development, 129
Teacher Training Institutes in Netherlands, 105, 106
Theoretically based competence models, 22
Transdisciplinarity, 36
Transformation, 185
Transformation competencies, 5, 13
Transformative education, 162
Transformative learning, 146
Transformative outdoor education (TOE), 168, 169, 173
Transformative pedagogies, 146
Transformative posture, 114
Transmissive 'banking' approach, 41

U
Uncomfortable knowledge, 65
Underlying components (UCs), 116
UNECE framework, 21–23
UNESCO, 50, 105
United Nations Economic Commission for Europe (UNECE), 13
Universitat Oberta de Catalunya (UOC), 80
University of Gloucestershire, 175
University of Vechta
 staff training at, 125–126
UOC's competence framework, 80–82
USSR, 47

V
Values, 38

W
Whole institution approach, 117
Whole School Approach (WSA), 106
Willing, 51
Willingness, 46

Y
Youth and Culture, 113

GPSR Compliance
The European Union's (EU) General Product Safety Regulation (GPSR) is a set of rules that requires consumer products to be safe and our obligations to ensure this.

If you have any concerns about our products, you can contact us on

ProductSafety@springernature.com

In case Publisher is established outside the EU, the EU authorized representative is:

Springer Nature Customer Service Center GmbH
Europaplatz 3
69115 Heidelberg, Germany

www.ingramcontent.com/pod-product-compliance
Ingram Content Group UK Ltd.
Pitfield, Milton Keynes, MK11 3LW, UK
UKHW052005051026
15624UKWH00003B/566
* 9 7 8 3 0 3 0 9 1 0 5 4 9 *